Production Engineering and Robust Control

Edited by Majid Tolouei-Rad, Pengzhong Li and Liang Luo

Published in London, United Kingdom

Meet the editors

Dr. Majid Tolouei-Rad is currently with the School of Engineering, Edith Cowan University, Perth, Australia. He has worked as an engineer for industries and later as an academic for reputable universities in Iran, Canada, and Australia. Dr. Tolouei-Rad has been involved in tertiary teaching and research for more than two decades during which he has educated dozens of engineers. He has more than eighty publications in the form of books, book chapters, and research articles. His research interests include additive manufacturing, robotics and automation, and advanced manufacturing systems and technologies.

Dr. Pengzhong Li is a professor at the School of Mechanical Engineering and a teaching director at the Sino-German College of Postgraduate Studies (CDHK), Tongji University, Shanghai, China. He received his Ph.D. in Mechanical Engineering from Tongji University in 2004. From 1995 to 2001, he was a manager in the Business and Warehousing Management Departments of Guilin Daewoo Bus Co., LTD. He is a Director of the Intelligent Manufacturing and Services Branch of the China Creative Studies Institute and the Director of the East China Branch of the National College Institute of Manufacturing Automation. Dr. Li's research interests include intelligent manufacturing systems and quality control technology in manufacturing. At present, he is committed to research on man-machine coordination in the application of artificial intelligence.

Dr. Liang Luo is an assistant professor at the Sino-German College of Postgraduate Studies (CDHK), Tongji University, Shanghai, China. His research interests include manufacturing informatics of intelligent manufacturing systems, intelligent evaluation and optimization of manufacturing systems, and quality management. He received his Ph.D. in Mechanical Engineering from Tongji University in 2019. He is a member of the Intelligent Manufacturing and Services Branch of the China Creative Studies Institute.

Contents

Preface

The world's population is increasing rapidly and thus so is the need for producing more and more products each year. This requires the utilization of innovative and more efficient production technologies and systems, and a proper application of robust control systems for controlling processes, machines, robots, packaging and part transfer systems, and so on.

This book presents the latest achievements of researchers and engineers working on different aspects of production engineering and control systems. There are two sections: Section 1 includes four chapters on production engineering and Section 2 includes four chapters on robust control. Chapter 1 introduces production engineering and classifies different activities within this domain. Chapter 2 explains the strategy used for the optimization of process parameters in abrasive waterjet machining as a non-conventional cutting process. Chapter 3 is on additive manufacturing where inkjet printing is used for the 3D printing of smart products. Chapter 4 covers production systems and discusses the requirements for product lifecycle management. Chapter 5 explains the hierarchical sliding mode control for a 2D ballbot. Chapter 6 describes the controller modeling for proportional–integral–derivative (PID) gain tuning against external disturbances with constrained internal parameter variation of the permanent magnet DC motor based on an optimization technique. Chapter 7 focuses on the role of uncertainty in the development of the model and explains how a multi-robot control system can reduce various uncertainties. Finally, Chapter 8 proposes the design of the hardware architecture of an improved direct torque control (DTC) for real-time implementation.

The editors are thankful to those who contributed their findings for publication in this book. They would also like to thank Blanka Gugic, Zrinka Tomicic, Jelena Germuth, and other dedicated staff at IntechOpen for providing continuous support that made the publication of this book possible.

Dr. Majid Tolouei-Rad
School of Engineering,
Edith Cowan University,
Perth, Australia

Dr. Pengzhong Li
School of Mechanical Engineering,
Tongji University,
Shanghai, P.R. China

Dr. Liang Luo
Sino-German College of Postgraduate Studies,
Tongji University,
Shanghai, P.R. China

Section 1

Production Engineering

Chapter 1

Introductory Chapter: Production Engineering

Majid Tolouei-Rad

1. Introduction

Production engineering is a broad term covering many activities involved in the production life cycle of industrial products. The term itself is interchangeably used with manufacturing engineering as manufacturing is the backbone of production engineering.

There have been many developments in the domain of production engineering in recent years as reported in the literature. These include developments of new methods in production methods and systems and optimization for improving productivity or maximizing profit. For example, drilling as the most used production process is still the subject of study for improvement by contemporary researchers, and many of these works can be found in the literature [1–8]. There are also reports on the improvements of conventional and nonconventional production processes [9–11], optimization of processes [6, 12, 13], and enhancement of properties [14]. In addition, there are many reports in the literature on the development of production systems including quality control, process planning, and production planning and control systems [15–23].

2. Production processes

As shown in **Figure 1**, production engineering is broadly covering production processes and production systems, and each of these includes various operations. Production processes alter the shape, geometry, and properties of the workpiece enabling it to perform its function. Processing operations may include the enhancement of properties of the workpiece by means of heat treatment operations and improving the quality of the workpiece surfaces using surface processing operations. There is a variety of manufacturing operations including conventional and nonconventional methods. Generally, conventional manufacturing processes are preferred as these are often more economical and the equipment needed is readily available. However, in some cases, the use of nonconventional manufacturing processes is inevitable despite imposing higher costs. For example, **Figure 2** shows a flat metal part with a complex external shape and many internal cut geometries. This part is cut from an 8 mm thickness AISI 304 L stainless steel plate. It is a relatively thick plate of hard-to-machine stainless steel material, making its production problematic using conventional methods. The use of a nonconventional method such as Abrasive Water Jet Machining (AWJM) makes this a relatively easy operation. This part is taken from a research work studying the optimization of AWJM process parameters for improving

IntechOpen

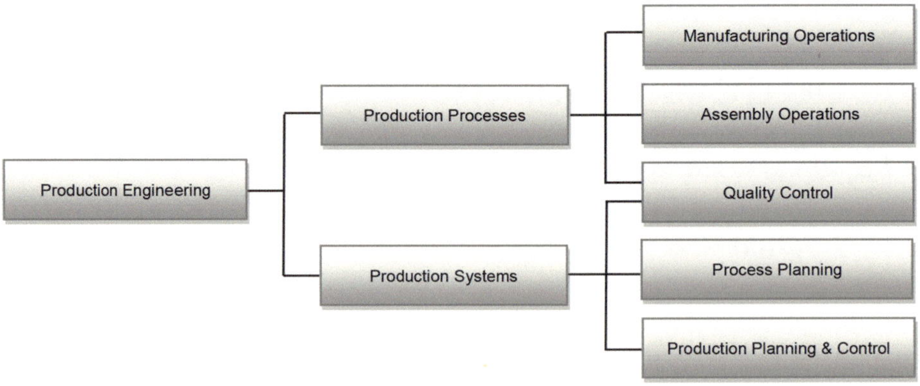

Figure 1.
Classification of activities in production engineering.

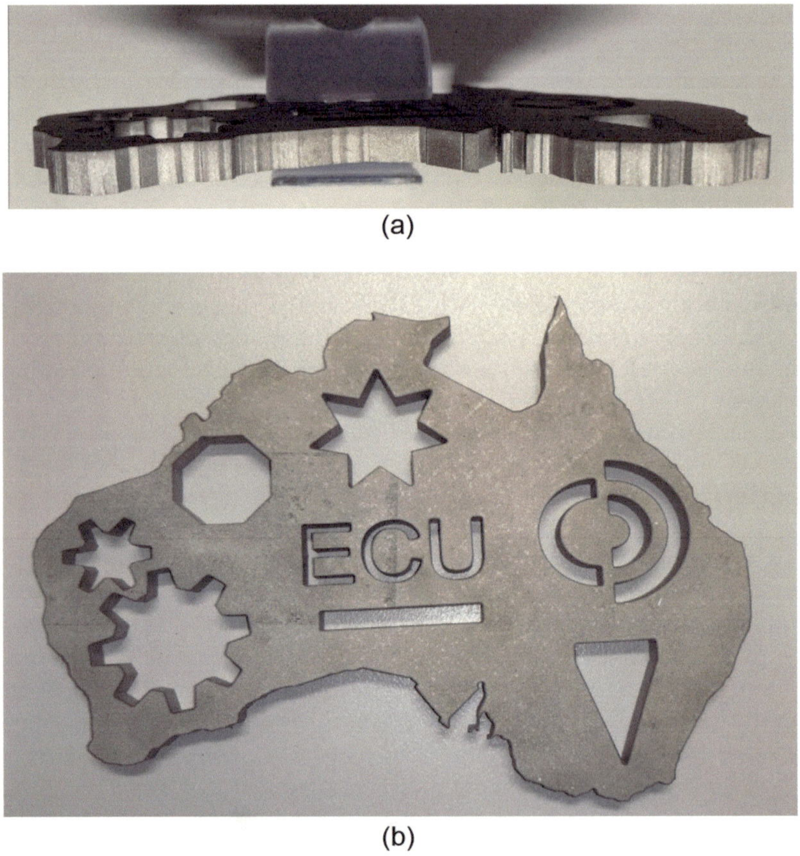

Figure 2.
Map of Australia with embedded geometrical shapes cut from an 8 mm thickness AISI 304 L stainless steel plate using abrasive water jet machining (AWJM). (a) Side view. (b) Top view.

productivity [24]. Another example is shown in **Figure 3** where a complex lattice structure is formed by connecting multiple curved surfaces with thin wall thicknesses. The material of this part is 316 stainless steel which is hard to machine, and

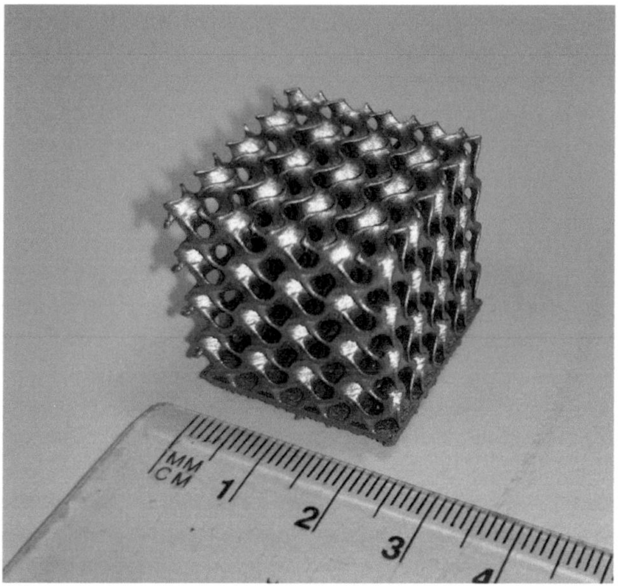

Figure 3.
A lattice structure printed from 316 stainless steel power using the selected laser melting (SLM) technique.

its complex geometry and small wall thickness make its production impossible with conventional methods. It is produced by additive manufacturing (AM) also known as 3D printing from a powder-like material. This part is printed layer by layer using a layer thickness of 25 micrometers where powder-like metal particles are melted using a powerful laser beam in a process known as selective laser melting (SLM). The process is costly, and the rate of production is low; however, making it possible to produce some complex geometries that would have not been impossible otherwise.

Cutlery items such as spoons and forks are examples of single-part products where no joining or assembly operations are needed after the single-part product is manufactured. However, most products consist of more than one part and there is a need for joining or assembling of these parts. The number of parts in a product where assembly or joining processes are needed can go from two to millions. A screwdriver has 2, a typical car has about 20,000, and the largest passenger aircraft has over 4 million parts that are individually manufactured and assembled to form a complex product with many assemblies and subassemblies. Many types of joining, and assembly operations are used in production plants. Some are permanent joining methods such as welding, brazing, soldering, riveting, and adhesives; and some provide the possibility of disassembly where mechanical fasters such as bolts and nuts or screws are used.

Quality control is often considered the last step of the production cycle although modern manufacturing strategies state that quality must be built into the product and must be incorporated during manufacturing and assembly operations. In either case, the quality of the product must be tested and verified before it leaves the production plant.

3. Production systems

In general, production processes are referred to those operations where there is a need for physical contact of the product with processing equipment, or the worker;

and a physical contact is not needed when the product is processed by production systems. Although some noncontact quality control methods are used, in most quality control activities there is a need for measuring the dimensional accuracy of the product or its properties requiring physical contact. Yet the quality control information is widely used in production systems for analysis of production where there is no need for physical contact. The assessment of product quality does not end in the production plant and it is still under quality assessment after delivering to the customer and beginning its service life. This is possible by receiving feedback from the customer and also by providing after-sales services. Accordingly, as shown in **Figure 1**, quality control is an activity that relates itself to both production processes and production systems.

In the literature, the production systems are also referred to as manufacturing support systems as processing operations cannot be accomplished without these activities. In addition to quality control, production systems include process planning and production planning and control systems. When a product is to be produced, one of the first steps is to design the product such that it meets the intended specifications. The number of parts, assemblies and subassemblies, processing operations, and equipment should all be identified in an activity known as process planning. Various methods of process planning and the level of detailed information provided on the process sheet varies in different production plants.

Production planning and control (PPC) takes into consideration the logistics problems such as how many products are to be produced in a day or in a year, how long the production line will continue producing the part or product considering market demands, what are the raw material and equipment requirements for responding to production needs, and so on. When production begins there is a need for production control to ensure that production is running smoothly, will be meeting the planned completion dates, and any potential problems that could disrupt a smooth production are identified and tackled.

Author details

Majid Tolouei-Rad
School of Engineering, Edith Cowan University, Joondalup, WA, Australia

*Address all correspondence to: m.rad@ecu.edu.au

References

[1] Sobri SA, Heinemann R, Whitehead D. Carbon fibre reinforced polymer (CFRP) composites: Machining aspects and opportunities for manufacturing industries. In: Composite Materials: Applications in Engineering, Biomedicine and Food Science. Cham: Springer International Publishing; 2020. pp. 35-65

[2] Aamir M, Tolouei-Rad M, Giasin K. Multi-spindle drilling of Al2024 alloy and the effect of TiAlN and TiSiN-coated carbide drills for productivity improvement. The International Journal of Advanced Manufacturing Technology. 2021;**114**(9):3047-3056

[3] Tolouei-Rad M, editor. Drilling Technology. London, UK: IntechOpen; 2021 [Online]. Available from: https://www.intechopen.com/books/10375. DOI: 10.5772/intechopen.91561

[4] Tolouei-Rad M, Aamir M. Introductory chapter: Drilling technology. In: Drilling Technology. London, UK: IntechOpen; 2021 [Online]. Available: https://www.intechopen.com/chapters/76535. DOI: 10.5772/intechopen.97648

[5] Tolouei-Rad M, Aamir M. Analysis of the performance of drilling operations for improving productivity. In: Drilling Technology. London, UK: IntechOpen; 2021 [Online]. Available: https://www.intechopen.com/chapters/75613. DOI: 10.5772/intechopen.96497

[6] Vafadar A, Hayward K, Tolouei-Rad M. Drilling reconfigurable machine tool selection and process parameters optimization as a function of product demand. Journal of Manufacturing Systems. 2017;**45**:58-69

[7] Tolouei-Rad M, Shah A. Development of a methodology for processing of drilling operations. International Journal of Industrial and Manufacturing Engineering. 2012;**6**(12):2660-2664

[8] Aamir M, Giasin K, Tolouei-Rad M, Ud Din I, Hanif MI, Kuklu U, Pimenov DY, Ikhlaq M. Effect of cutting parameters and tool geometry on the performance analysis of one-shot drilling process of AA2024-T3. Metals. 2021;**11**(6):854

[9] Sahoo SP, Pandey K, Datta S. Performance of uncoated/coated carbide inserts during MQL (sunflower oil) assisted machining of Inconel 718 superalloy. Sādhanā. 2022;**47**(4):1-31

[10] Jalali Azizpour M, Tolouei-Rad M. Evaluation of residual stress in HVOF stellite-6 coatings using non-contact drilling. Materials Research Express. 2019;**6**(6):066577

[11] Xu M, Wei R, Li C, Kurniawan R, Chen J, Ko TJ. Comprehensive study on the cutting force modeling and machinability of high frequency electrical discharge assisted milling process using a novel tool. International Journal of Precision Engineering and Manufacturing-Green Technology. 2022;**15**:1-28

[12] Aamir M, Tu S, Tolouei-Rad M, Giasin K, Vafadar A. Optimization and modeling of process parameters in multi-hole simultaneous drilling using taguchi method and fuzzy logic approach. Materials. 2020;**13**(3):680

[13] Tolouei-Rad M, Bidhendi IM. On the optimization of machining parameters for milling operations. International Journal of Machine Tools and Manufacture. 1997;**37**(1):1-6

[14] Tolouei-Rad M, Lichter E. The heat treatment analysis of e110 case hardening steel. Journal of Engineering Science and Technology. 2016;**11**(3):407-415

[15] Muthiah KM, Huang SH. A review of literature on manufacturing systems productivity measurement and improvement. International Journal of Industrial and Systems Engineering. 2006;**1**(4):461-484

[16] Tolouei-Rad M. Integration of part classification, cell formation and capacity adjustment. Journal of Achievements in Materials and Manufacturing Engineering. 2010;**39**(2):197-203

[17] Tamás P, Illés B. Process improvement trends for manufacturing systems in industry 4.0. Academic Journal of Manufacturing Engineering. 2016;**14**(4):119-125

[18] Tolouei-Rad M. Intelligent analysis of utilization of special purpose Machines for Drilling Operations. In: Intelligent Systems. London, UK: IntechOpen; 2012 [Online]. Available: https://www.intechopen.com/chapters/30663. DOI: 10.5772/36896

[19] Zhao C, Li J, Huang N. Efficient algorithms for analysis and improvement of flexible manufacturing systems. IEEE Transactions on Automation Science and Engineering. 2015;**13**(1):105-121

[20] Tolouei-Rad M. An efficient algorithm for automatic machining sequence planning in milling operations. International Journal of Production Research. 2003;**41**(17):4115-4131

[21] Robinson A. Modern Approaches to Manufacturing Improvement: The Shingo System: The Shingo System. New York, USA: Routledge; 2017

[22] Mula J, Poler R, García-Sabater JP, Lario FC. Models for production planning under uncertainty: A review. International Journal of Production Economics. 2006;**103**(1):271-285

[23] Kang CW, Ramzan MB, Sarkar B, Imran M. Effect of inspection performance in smart manufacturing system based on human quality control system. The International Journal of Advanced Manufacturing Technology. 2018;**94**(9):4351-4364

[24] Llanto JM, Vafadar A, Tolouei-Rad M. Multi-objective optimisation in abrasive waterjet contour cutting of AISI 304L. In: Production Engineering [Working Title]. London, UK: IntechOpen; 2022 [Online]. Available: https://www.intechopen.com/online-first/83563. DOI: 10.5772/intechopen.106817

Chapter 2

Multi-objective Optimisation in Abrasive Waterjet Contour Cutting of AISI 304L

Jennifer Milaor Llanto, Ana Vafadar and Majid Tolouei-Rad

Abstract

The optimum waterjet machining parameters were found for maximising material removal rate and minimising surface roughness and kerf taper angle where three levels of traverse speed, abrasive flow rate, and waterjet pressure are used. The multi-linear regression equations were obtained to investigate the relationships between variables and responses, and the statistical significance of contour cutting parameters was analysed using the analysis of variance (ANOVA). Further, the response surface methodology (desirability function approach) was utilised for multi-objective optimisation. The optimum traverse speeds were 95 mm/min for 4 mm thickness and 90 mm/min for both 8 and 12 mm thicknesses. For all material thicknesses, the abrasive mass flow rate and waterjet pressure were 500 g/min and 200 MPa, respectively. The minimum values of surface roughness, kerf taper angle, and maximum material removal rate for 4-, 8- and 12-mm material thicknesses were respectively 0.799°, 1.283 µm and 297.98 mm^3/min; 1.068°, 1.694 µm and 514.97 mm^3/min; and 1.448°, 1.975 µm and 667.07 mm^3/min. In this study, surface roughness and kerf taper angle decreased as the waterjet pressure and abrasive mass flow rate increased; and this is showing a direct proportional relationship with traverse speed, abrasive mass flow rate and waterjet pressure.

Keywords: abrasive water jet, contour cutting, surface roughness, kerf taper angle, material removal rate, response surface methodology, multi-objective optimisation

1. Introduction

Contour cutting is one of the processes applied in metal fabrication industries. There are several non-traditional technologies employed for contour cutting, such as electro discharge machining, laser beam machining and electrochemical discharge machining, that have been noted to provide exemplary performance [1]. Accordingly, Abrasive Water Jet Machining (AWJM) is an advanced manufacturing techniques that demonstrated advantages to non-traditional machining technology owing to: its capability in cutting complex geometries, its absence of tool wear, its absence of thermal distortion, and it being environmentally friendly [2, 3]. The cutting process in AWJM is based on removing materials from a target workpiece via erosion [4]. Within this process, contour profiles in various types of programs are downloaded in a computer-based controller, where subsequently a high-pressure pump releases pressurised water in the nozzle system. The pressurised water, moving with a high velocity, is released from the orifice in a very thin stream structure [5]. The high-speed water jet that contains abrasive particles is then accelerated to generate an abrasive waterjet. Finally, the focusing tube drives the abrasive waterjet to its target point for cutting the material [4, 6]. The compounded granular abrasive and high-pressure waterjet stream makes the abrasive waterjet capable of machining various workpieces, such as metals.

The performance of AWJM is influenced by several process parameters, which can be varied constantly within a period. In general, the primary goal of the metal fabrication industry is to manufacture high quality products in a shortened period. To attain productivity and economy objectives, it is imperative to select an optimum combination of process parameters within the abrasive waterjet cutting processes. Conventionally, the identification of the most suitable values of process parameters is accomplished by the execution of many experiments. Hence, to establish the optimum combination of process parameters in the absence of extensive experimental exertion, researchers have utilised advanced modelling techniques and optimisation in progressing the performance of abrasive waterjet cutting. For instance, Rao et al. [7] have investigated the impacts of traverse speed, standoff distance and abrasive mass flow rate in AWJM of AA631-T6. They have considered single-objective and multi-objective optimisation attributes to achieve optimum solutions by utilising Jaya and MO-Jaya algorithms, which were a posterior optimisation used to solve constrained and unconstrained conditions. The objectives of maximising material removal and minimising kerf taper angle and surface roughness were achieved by lower traverse speed and standoff distance and higher abrasive mass flow rate. Moreover, they determined that multi-objective Jaya algorithm achieved better results as compared with other algorithms, such as simulated annealing (SA), particle swam optimization (PSO), firefly algorithm (FA), cuckoo search (CS) algorithm, blackhole (BH) algorithm, bio-geography-based optimization (BBO) algorithm, non-dominated sorting genetic algorithm (NSGA), non-dominated sorting teaching-learning-based optimization (NSTLBO) algorithm and sequential approximation optimization (SAQ). Nair and Kumanan [8] have similarly applied weighted principal components analysis (WPCA) for optimising AWJM process parameters in machining Inconel 617. These authors evaluated the impacts of abrasive mass flow rate, standoff distance, table feed and waterjet pressure against material removal rate and geometric accuracy. The WPCA method uses internal tests and training samples to calculate the 'weighted' covariance matrix, establishing that an increase in standoff distance enhances the abrasive flow volume, leading to less geometric errors and a higher rate of material removal. Equivalently, Chakraborty and Mitra [9] have applied the grey wolf optimiser (GWO) technique for AWJM cutting of AL6061to maximise material removal rate and minimise surface roughness, simultaneously considering the constrained values of input parameters i.e., nozzle diameter and titled angle, jet feed speed, surface speed, waterjet pressure and abrasive mass flow rate. This algorithm demonstrated a faster hunting of prey (discovering the optimum parameter settings), due to the existence of a social hierarchy of grey wolves. They achieved maximum MRR via higher rate of nozzle titled angle, surface speed, waterjet pressure and abrasive mass flow rate. In the case of surface roughness, it attained its minimum value at lower rate of waterjet pressure, jet feed and surface speed and higher rate of abrasive mass flow. Trivedi et al. [10] have examined the impacts of process parameters such as pressure, traverse rate and standoff distance on surface integrity in AWJM of AISI 316 L. Analysis of variance was employed to develop an empirical model by regression analysis for surface roughness. These authors concluded traverse speed to be the most significant parameter influencing surface roughness, whereby increasing pressure improved the surface quality of the target workpiece. Additionally, they established standoff distances, as the least contributing parameter. Research focused on optimisation of cutting operations is being continuously undertaken by researchers, where varied methods have been employed to solve different single and multi-objective optimisation problems [11–14]. Whereas single-objective optimisation problems have conventionally been applied, the performance of AWJM has mainly been measured based on multiple responses. In accordance, a multi-objective approach is required in order to optimise several categories of objective functions simultaneously. Several methods have been developed to date, and are continuously being progressed, in order to solve single-objective problems. Advances in optimisation techniques, such as: genetic algorithms (GA), simulated annealing (SA), artificial bee colony (ABC), ant colony optimization (ACO), particle swarm optimization (PSO) and teaching-learning-based optimization (TLBO), and others, have been demonstrated to be remarkably efficient in defining the optimum value of AWJM process parameters [15].

In abrasive waterjet contour-cutting, it has been realised that the impacts of most influencing factors, such as waterjet pressure, abrasive mass flow rate, standoff distance and traverse speed in straight-slit cutting, are similar with contour cutting. These research studies have shown the application of computational approaches for optimising process parameters in abrasive waterjet contour cutting requires further investigation. Therefore, this research considers the optimisation of relevant process parameters, including traverse speed, abrasive mass flow rate, and waterjet pressure on surface roughness, material removal rate and kerf taper angle in abrasive waterjet contour cutting of AISI 304L of varied thicknesses.

In this work, the experiment was designed using Taguchi orthogonal array, where a regression model has been developed to formulate the optimisation fitness function. This modelling technique has been applied to predict the response and determine optimum process parameters. In addition, response surface methodology (RSM) has been employed for multi-objective optimization, in order to obtain optimum values of input process parameters and to investigate the impacts and interactions against response parameters.

2. Methodology

In this study, three major steps were employed, consisting of abrasive waterjet contour cutting experiments, regression modelling and optimisation. The experiment, modelling and optimisation procedures are presented in **Figure 1**. The experiment was conducted using the Taguchi L_9 orthogonal array to analyse the impacts of input parameters, i.e., traverse speed, abrasive mass flow rate and waterjet pressure. Desirability analysis using response surface methodology is employed for the experimental results of material AISI 304L. In this desirability analysis, multi-responses are considered. It establishes the optimum set of the selected process parameters on the performance characteristics.

A regression model was developed using the machining process parameters from the experimental execution to extract mathematical models. A linear stepwise regression analysis was performed to predict the surface roughness, material removal rate

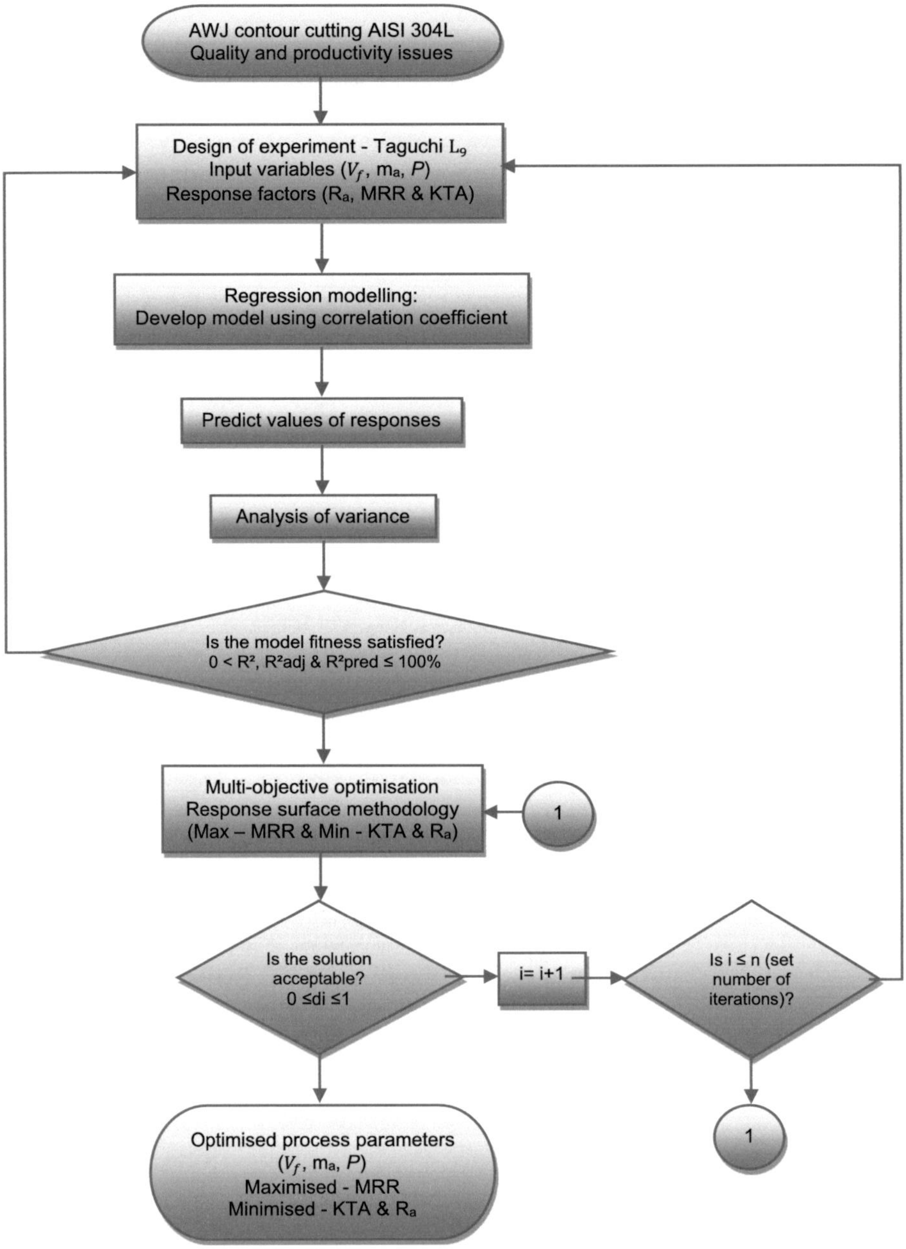

Figure 1.
Multi-objective optimisation process flow chart.

and kerf taper angle value. The reliability of the models generated was assessed based on coefficient of determination (R^2, R^2adj & R^2pred). However, supposing that regression models are not within the acceptable range or do not provide preferable values of coefficients of determination set by the decision-maker, it is anticipated that these models will not provide precise prediction. Therefore, the selected parameter setting conflicts with the response variables, denoting the necessity for modification of independent variables or experimental design [16].

Referring to **Figure 1**, after achieving the fittest models, a multi-objective optimisation was performed by using response surface methodology with the objectives of maximising material removal, whilst minimising surface roughness and kerf taper angle. The number of solutions and iterations (i = 1 to n) may vary, depending on the machining process requirements to establish the best alternative or solution. Hence, if the composite desirability is not within the tolerable array, several iterations repeating the response surface optimisation were executed. Subsequently, if these repetitions reached the maximum number of iterations and the composite desirability is not attaining adequate values, modifying the design of experiments and the corresponding independent variables or its values is necessary [16]. Moreover, in some cases, other soft computing techniques should be considered [17].

2.1 Material and experimental design

In this work, the material machined in the experiments was AISI 304L with varied thicknesses of 4, 8 and 12 mm. The assigned material thicknesses with differing uniform gaps were used to gain a better yield of variations in AWJM cutting behaviour. Stainless steel, such as AISI 304L, is widely used in fabrication industries, where it is recognised for its high strength and corrosion and heat resistance. This results from its high alloying content of Cr and Ni [18]. The chemical and mechanical composition of this material is detailed in **Table 1**.

The setup consisted of an OMAX MAXIEM 1515 abrasive waterjet machine, possessing a direct drive pump and dynamic cutting head with maximum pressure of 413.7 MPa and cutting area of 2235 mm length and 1727 mm width. The cutting head is comprised of a mixing chamber for abrasive and waterjet, along with a nozzle diameter of 0.56 mm and a jet impact angle of 90°. An abrasive garnet with a mesh size of #80 was utilised for abrasive waterjet cutting experiments. The unit is inclusive of IntelliMax software, where the experiment setup conditions were uploaded and entered. The cutting head can move in the Z-axis over a distance of 305 mm, with a maximum traverse speed of 12,700 mm/min. Standoff distance was designated to 1.5 mm in agreement with recommended range for abrasive waterjet machining in previous works [20, 21]. The AWJM setup and process parameters are demonstrated in **Figure 2**.

Chemical composition in wt.%		Mechanical properties	
C	**0.03**	**Hardness, Rockwell B**	82
Mn	2	Tensile Strength, Ultimate, MPa	564
Si	0.75	Tensile Strength, Yield, MPa	210
Cr	18.00–20.00	Elongation at Break	58%
Ni	8.00–12.00	Modulus of Elasticity, GPa	193–200
P	0.045		
S	0.03		
Ni	0.1		
Fe	Remaining		

Table 1.
Chemical and mechanical composition of AISI 304L [19].

Upon completion of the experiments, the roughness of the machined surfaces was quantified by a surface roughness tester (TR200 model). **Figure 2** presents the cut surface captured by LEICA M80, which indicates the measurement area for the roughness. The kerf top and bottom width were measured using a LEICA M80 optical microscope model. Moreover, rate of material removal and kerf taper angle were calculated using Eqs. (1) and (2), respectively [11]. The roughness of the cut surface determined according to the ISO/TC 44 N 1770 standard, (μm); W_t is width of the cut surface at the jet inlet, (mm]; W_b is the width of the cut surface at the jet outlet, (mm); u is the angularity or perpendicular deviation, (mm); α°- inclination angle of the cut surface, (°); MRR is the Material Removal Rate, (mm^3/min); t is the thickness of the material (mm) [22].

$$MRR = ht\left(\frac{W_t + W_b}{2}\right)V_f \tag{1}$$

$$KTA = Arctan\left(\frac{W_t + W_b}{2ht}\right) \tag{2}$$

The input parameters considered in abrasive waterjet contour cutting in this experiment included traverse rate (V_f), abrasive flow rate (m_a) and water pressure (P), as these parameters have been demonstrated in previous studies as having significant impacts in AWJM applications [10, 12, 23, 24]. Surface integrity, kerf geometries

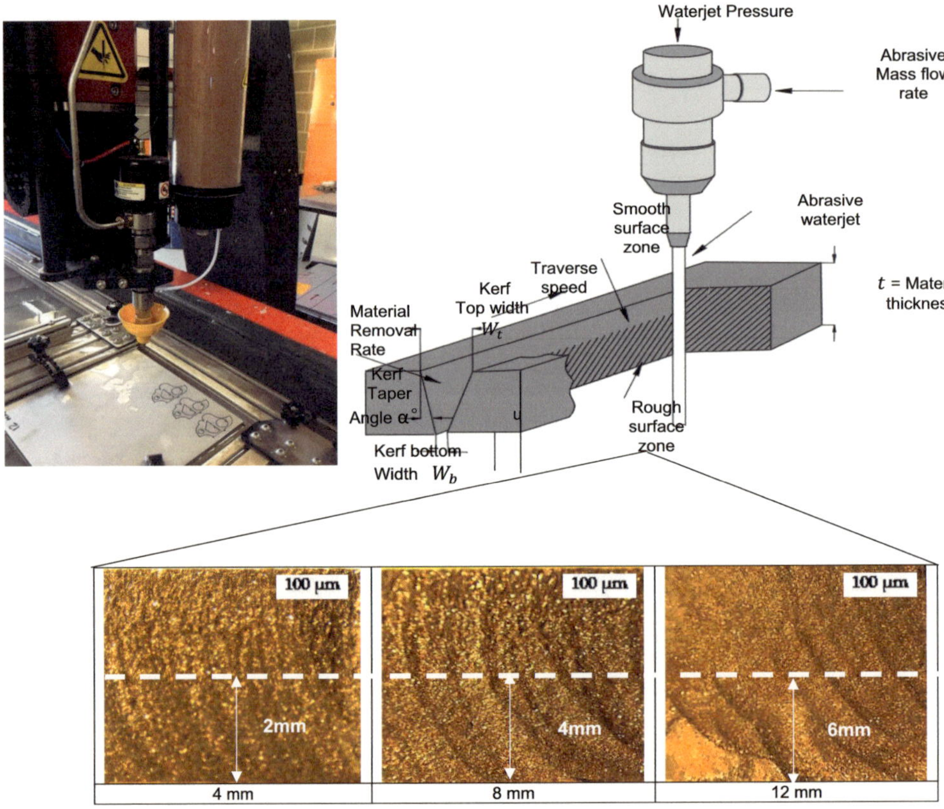

Figure 2.
AWJM setup and process parameters.

and low material removal rate evidence has been reported in machining of AISI 304L, requiring further improvement [4, 25]. Furthermore, taper angles formed in AWJM demonstrate different inclinations as contour curvature radius differs [26]. Hence, quality and productivity are an intensified demand in various manufacturing fields and are significant performance indicators for machining processes. Therefore, in this study, material removal rates (MRR), surface roughness (R_a) and kerf taper angle (KTA) have been chosen as process parameter characteristics for abrasive waterjet contour cutting investigations, due to their influence against the selected input parameters. The levels of the considered independent variables, responses and coding assignment have been detailed in **Tables 2** and **3**.

Abrasive waterjet cutting was executed for three different profiles, representing straight-line, inner arcs and outer arcs, as part of the completed twelve profiles, as demonstrated in **Figure 2**. The abovementioned profiles were selected to confirm a broad array of complicated machining profiling applications. The levels of profiles employed showed occurrences of surface roughness, low machining rate and inaccuracies of cut geometries in regard to previous works [27, 28], recommending further studies, predominantly for difficult-to-cut materials, such as AISI 304L (**Figure 3**).

The design of experimentation (DOE) was carried out using the Taguchi approach in MINITAB 19 software. The Taguchi method is useful in determining the best combination of factors under desired experimental conditions, reducing the large number of experiments which would be required in traditional experiments as the number of process parameter increases [29, 30].

In Taguchi's approach, selection of the appropriate orthogonal array depends on aspects such as: the number of input and response factors along with the interactions that are of key significance; number of levels of data for input factors; and required resolution of experiment and limitations cited on cost and performance [29, 31]. With this specific advantage, this method is suitable in conducting experiments with an appropriate number of tests to determine the optimal combination and significance of the selected factors [32]. The relevant variation in thicknesses dictates different material responses. Therefore, Taguchi L_9 orthogonal array was executed for three

Independent variables	**Codes**	**Levels**		
		1	**2**	**3**
Traverse speed, V_f = mm/min	X_1	90	120	150
Abrasive mass flow rate, m_a = g/min	X_2	300	400	500
Waterjet pressure, P = MPa	X_3	200	250	300

Table 2.
Levels of input process parameters.

Profiles	**Surface roughness, μm**	**Material removal rate, mm^3/min**	**Kerf taper angle, ^{0}**
Straight-line, 20 mm	R_a1	MRR1	KTA1
Inner arc, R10	R_a2	MRR2	KTA2
Outer arc, R20	R_a3	MRR3	KTA3

Table 3.
Output parameters for varied profiles.

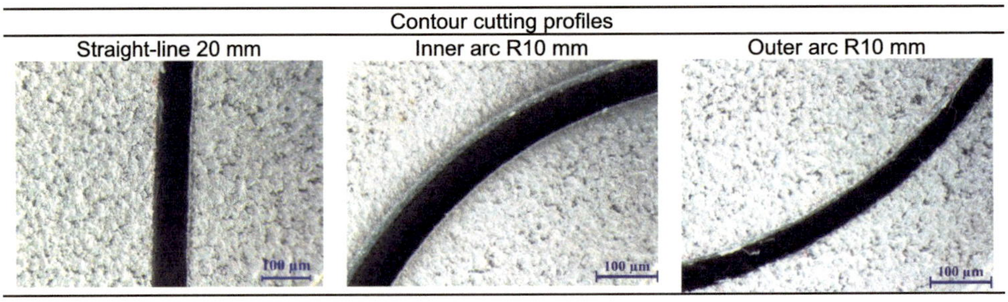

Figure 3.
Abrasive waterjet contour cutting profiles.

Exp. No.	Input Parameters		
	V_f	m_a	P
	(mm/min)	(g/min)	(MPa)
1	90	300	200
2	90	400	250
3	90	500	300
4	120	300	250
5	120	400	300
6	120	500	200
7	150	300	300
8	150	400	200
9	150	500	250

Table 4.
Taguchi L_9 *orthogonal array.*

levels of material thicknesses (t), i.e., 4, 8 and 12 mm, as presented in **Table 4**. The AWJM performances were analysed accordingly by the applied material thickness.

2.2 Modelling and multi-objective optimisation

A mathematical model was developed to associate the input process parameters to the response's characteristics. To achieve this, a linear regression was employed to develop models for the prediction of responses. The empirical model for the prediction of the responses in regard to controlling parameters was established by linear regression analysis. Regression analysis was then applied to obtain the interactions between independent and dependent variables [33]. Multi-linear regression involves regression analysis of dependent and independent variables exhibiting a linear relationship [34]. It stipulates the relationship between two or more variables and a response variable by fitting a linear equation to examine data. The value of the independent variable x or process parameter is correlated with a value of the dependent variable, y, which is the output parameter. In general, this analysis is applied to investigate the degree of relationship between multiple variables fitted by a straight line [33].

In general, regression model is expressed by Eq. (3) [33].

$$y = \propto + \beta_1 x_1 + e \tag{3}$$

$$\text{Wherein} : e = y_1 - \widehat{y_1} \tag{4}$$

where, y = dependent variable, α = constant, x_1 = Independent variable, β_1= coefficient of independent variablex_1, e = error, y_1= regression line values and $\widehat{y_1}$ = actual observation.

If this involves more than one variables, then it is categorised as multi-regression as shown in Eq. (5) [33].

$$y = \propto + \beta_1 x_1 + \beta_2 x_2 + \beta_3 x_3 + \dots \dots \dots \beta_n x_n + e_n \tag{5}$$

A multi-linear regression analysis can be employed *to fit a predictive model to an observed data set of values of output and input variables*. The obtained results of surface roughness, material removal rate and kerf taper angle were expressed in terms of the input parameters such as traverse speed (X_1) abrasive mass flow rate (X_2) and waterjet pressure (X_3).

The predicted values are functional for optimising the parameters by providing an adequate comprehension of the significant parameters. The percentage of error between the experimental data and acquired predicted values has been calculated based on Eq. (6) [33]. The relative percentage of error was acceptable at <20% [35].

$$Error = \frac{1}{n} \sum_{n}^{1} \left[\frac{Response_{(experiment)} - Response_{(predicted)}}{Response_{(experiment)}} \right] \% \tag{6}$$

The performance of the established regression model was assessed by statistical approaches to confirm the goodness-of-fit of the model and the impact of the predicted variables. Following this, the significance and effectiveness of the developed models were validated by *analysis of variance. Analysis of variance (ANOVA) is a statistical method that facilitates the evaluation of comparative influences for each control parameter* [36, 37]. The significance of input parameters including traverse speed, abrasive mass flow rate and waterjet pressure were investigated using p- values and determination of coefficient (R^2). In this work, a confidence interval of 95% (p < 0.05) has been applied that is in alignment with previous works [29, 38, 39]. A 95% confidence interval means that there is only a 5% chance of being the wrong estimation; therefore, the influence of each process parameter or other interactions on the responses is considered insignificant if their p-values were estimated at more than 0.05 [37].

The determination of coefficient (R^2, R^2adj and R^2pred) refers to the percentage variation of responses ranging from 0–100%. These indicators determine the adequacy of the model against obtained experimental data and predicted observation. This R^2, R^2adj and R^2pred value of ≥80%, proved a better model fits of the obtained data [35].

Response surface methodology (RSM) can be utilised for multi-objective optimisation. This multi-desirability is based on multi-response optimisation using an objective function D(X), denoted as desirability function [40]. This method translates each response (y_i) into a desirability function (di), differing in the array of $0 \leq di \leq 1$, where desirability function =0 indicates an undesirable response and desirability function =1 represents a fully desired response [41]. The objective function D is specified by Eq. (7) [40].

$$D = (d_1 X d_2 X \ldots \ldots d_n)^{1/n} = \left(\prod_{i=1}^{n} d_i \right)^{1/n} \quad (7)$$

The effectiveness of multi-objective optimisation is anticipated based on the method used for establishing priority weights for each response characteristics [42]. Generally, equal importance is set for selected responses; hence, weights may differ depending on the machining process requirements in order to establish the most suitable solution [43].

A simultaneous optimisation process was employed to determine the levels of resulting to the maximum overall desirability. The responses namely R_a, MRR and KTA were optimised concurrently to assess the set of input process parameters with the objectives of maximising MRR and minimising R_a and KTA.

3. Results and discussion

3.1 Regression models and analysis for surface roughness

The multi-linear regression coefficients are summarised in **Table 5**, exhibiting the correlation between the input parameters and the output surface roughness for straight-line, inner and outer arc profiles for material thicknesses of 4, 8 and 12 mm. The values of coefficients for all profiles and thicknesses demonstrate a similar trend, showing that constant and variable X_1 is positive and variables X_2 and X_3 are negative. The coefficient indicates the change in the mean response relating in the variation of the specific term, whilst the other term in the model remains constant. The relationship between a term and response is denoted by the sign of the coefficient [44]. The negative correlation coefficient denotes an inverse relationship between variables and responses; and therefore, if it is positive as the coefficient increases, the response mean value also increases. Therefore, an increasing rate of traverse speed (X_1) results in an incremental value of surface roughness. Moreover, an increasing rate of abrasive mass flow and waterjet pressure indicates/obtains a decreasing value of surface roughness. The values of R^2, R^2adj and R^2pred for 4, 8 and 12 mm ranged from 94.33–99.08%, 90.94–98.52% and 88.66–96.17%, respectively. This indicates that regression models denote an acceptable confirmation of the relationship between the independent variables and R_a response, which denotes a high significance of the model. Therefore, the multi-linear model is reliable and can be utilised in the optimisation of process parameters. It can be observed that the R^2, R^2adj and R^2pred obtained from straight-line, inner and outer arcs profiles have a uniform gap of at least 2%, which is comparable for all material thicknesses. Hence, this minimal gap denotes an insignificant difference between the surface roughness achieved from straight and curvature profiles [36].

The results detailed in **Table 5** show that the highest value of R^2, R^2adj and R^2pred for 4 and 8 mm material thickness are achieved in R_a3 with the values of 97.26, 94.84 and 92.45%; 98.64, 97.82 and 95.06%; 99.08, 98.52 and 96.17% respectively. Thus, R_a2 achieved the highest percentage of R^2, R^2adj and R^2pred for 12 mm material thickness

Term	t = 4 mm			t = 8 mm			t = 12 mm		
	R_a1	R_a2	R_a3	R_a1	R_a2	R_a3	R_a1	R_a2	R_a3
	Coef	**Coef**	**Coef**	**Coef**	**Coef**	**Coef**	**Coef**	**Coef**	**Coef**
	1.418	1.5394	1.4256	2.097	1.8107	1.76	2.542	2.3854	2.272
β_1	0.003522	0.002944	0.003222	0.009814	0.003483	0.008869	0.005389	0.004276	0.003090
β_2	- 0.000310	- 0.000300	- 0.000217	- 0.001464	- 0.000422	- 0.000577	- 0.000450	- 0.000446	- 0.000515
β_3	- 0.001500	- 0.001300	- 0.001133	- 0.001955	- 0.000977	- 0.001920	- 0.002567	- 0.001924	- 0.001081
Model Summary									
R^2	95.26%	96.77%	97.26%	98.01%	98.16%	98.64%	97.73%	99.08%	94.33%
R^2 (adj)	92.41%	92.41%	94.84%	96.82%	97.05%	97.82%	96.37%	98.52%	90.94%
R^2 (pred)	90.58%	90.58%	92.45%	93.84%	93.77%	95.06%	93.33%	96.17%	88.66%

Table 5.
Summary of multi-linear regression coefficients for R_a.

with the values of 99.08%, 98.52% and 96.17% accordingly. Therefore, the most fitted and predominant models were R_a3 for both 4 and 8 mm, and R_a2 for 12 mm material thickness. The predicted R_a values of regression models applied for straight-line, inner and outer arcs profiles of three levels of material thicknesses are detailed in **Tables 6-8**. The percentage error obtained for 4, 8 and 12 mm AISI 304L thicknesses ranged from −4.22 to 3.44%, 3.30 to 6.71% and − 5.75 to 2.49%, respectively. The errors determined for R_a between the predicted value and experimental results are less than 20%, denoting that these models are reliable for predicting R_a values.

Figure 4 presents the residual plot for R_a, consisting of normal probability plot, residual versus fits, histogram for residuals and residuals versus experimental values for the most fitted regression models for 4, 8 and 12 mm, at R_a3, R_a3 and R_a2, respectively. Similarly, the normal probability plots for all the material thicknesses demonstrated a close fit to a line in a normal probability graph. The points forming an

Exp. no.	Independent variables			R_a1 (μm)			R_a2 (μm)			R_a3 (μm)		
	X_1	X_2	X_3	Exp.	Pred.	Error %	Exp.	Pred.	Error %	Exp.	Pred.	Error %
1	90	300	200	1.35	1.34	0.81	1.43	1.42	0.61	1.46	1.45	0.56
2	90	400	250	1.25	1.24	1.41	1.33	1.35	−1.56	1.37	1.36	1.06
3	90	500	300	1.09	1.13	−3.99	1.25	1.27	−1.72	1.24	1.26	−2.44
4	120	300	250	1.36	1.37	−1.26	1.46	1.46	−0.39	1.48	1.48	0.22
5	120	400	300	1.29	1.27	2.35	1.42	1.39	3.44	1.40	1.38	1.72
6	120	500	200	1.41	1.39	2.45	1.50	1.48	2.28	1.48	1.48	−0.28
7	150	300	300	1.41	1.40	0.68	1.50	1.50	−0.39	1.49	1.50	−1.11
8	150	400	200	1.48	1.52	−4.22	1.58	1.60	−1.56	1.58	1.60	−2.11
9	150	500	250	1.43	1.42	1.77	1.51	1.52	−0.72	1.53	1.51	2.39

Table 6.
Predicted R_a values of regression models for t = 4 mm.

Exp. no.	Independent variables			R_a1 (μm)			R_a2 (μm)			R_a3 (μm)		
	X_1	X_2	X_3	Exp.	Pred.	Error %	Exp.	Pred.	Error %	Exp.	Pred.	Error %
1	90	300	200	2.12	2.15	−3.30	1.81	1.80	1.06	2.01	2.00	0.47
2	90	400	250	1.88	1.91	−2.73	1.72	1.71	0.86	1.86	1.85	0.98
3	90	500	300	1.64	1.66	−2.16	1.60	1.62	−2.04	1.65	1.69	−4.40
4	120	300	250	2.41	2.35	6.71	1.84	1.86	−1.81	2.16	2.17	−1.14
5	120	400	300	2.14	2.10	3.38	1.78	1.77	1.29	2.08	2.02	6.23
6	120	500	200	2.22	2.15	6.30	1.83	1.82	0.74	2.16	2.15	0.79
7	150	300	300	2.52	2.54	−1.89	1.92	1.91	0.63	2.32	2.34	−2.15
8	150	400	200	2.56	2.59	−3.68	1.95	1.97	−1.92	2.46	2.48	−1.58
9	150	500	250	2.32	2.35	−2.62	1.89	1.88	1.18	2.33	2.32	0.79

Table 7.
Predicted R_a values of regression models for t = 8 mm.

Exp. no.	Independent variables			R_a1 (μm)			R_a2 (μm)			R_a3 (μm)		
	X_1	X_2	X_3	Exp.	Pred.	Error %	Exp.	Pred.	Error %	Exp.	Pred.	Error %
1	90	300	200	2.39	2.38	1.17	2.26	2.25	0.42	2.18	2.18	−0.09
2	90	400	250	2.20	2.21	−0.50	2.12	2.11	0.60	2.09	2.07	1.20
3	90	500	300	2.00	2.03	−3.17	1.95	1.97	−1.98	1.99	1.97	2.49
4	120	300	250	2.42	2.41	0.83	2.29	2.28	0.30	2.22	2.22	0.69
5	120	400	300	2.25	2.24	1.17	2.16	2.14	2.03	2.05	2.11	−5.75
6	120	500	200	2.48	2.45	3.00	2.29	2.29	−0.39	2.15	2.17	−2.15
7	150	300	300	2.45	2.45	0.50	2.30	2.32	−1.36	2.27	2.26	1.46
8	150	400	200	2.60	2.66	−5.67	2.45	2.46	−1.34	2.32	2.31	0.43
9	150	500	250	2.51	2.48	2.67	2.34	2.32	1.74	2.22	2.21	1.72

Table 8.
Predicted R_a values of regression models for t = 12 mm.

approximately straight-line and falling along the fitted line denotes that the data is normally distributed and there is a good relation between measured and estimated response values [45]. In general, the residuals versus fits and observation graph for each material thickness display that the points are distributed randomly and near both sides of 0, with no distinguished pattern denoting a minimal deviation within residuals and estimated values. This graph plots the difference between the experimental data as predicted on the y-axis and the fitted or predicted values on the x-axis, to validate the assumption that the residuals have constant variance [46].

Figure 4 also exhibits the histogram graph for R_a, illustrating the distribution or frequency of the residuals for all observations. The data shows the frequency of R_a for 4, 8 and 12 mm material thicknesses to range from −0.02 to 0.03, −0.05 to 0.05 and − 0.02 to 0.02, respectively. The histogram presents distribution of the surface roughness obtained from varying material thicknesses. **Figure 4** histogram of residuals denotes that the residuals are normally distributed. These results reveal a minimal interval of inequalities of the experimental data, indicating that the R_a models meet their assumptions and are well fitted for the accuracy of prediction [46]. The effects of process parameters were established by ANOVA, where surface roughness results are given in **Tables A1-A3** in the Appendix section.

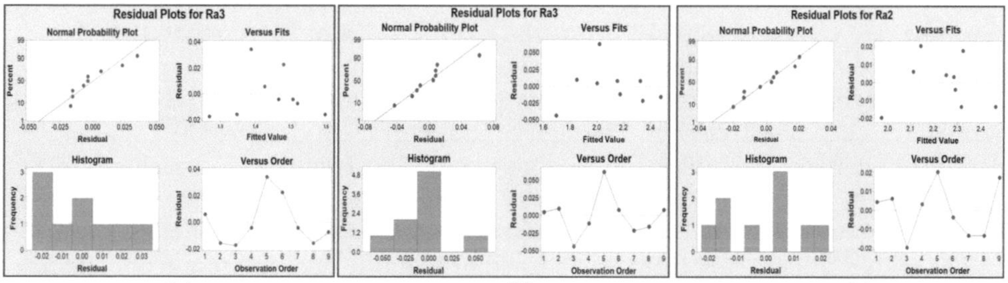

(a) (b) (c)

Figure 4.
Residual plots for surface roughness. (a) R_a3 (μm) for t = 4 mm (b) R_a3 (μm) for t = 8 mm (c) R_a2 (μm) for t = 12 mm.

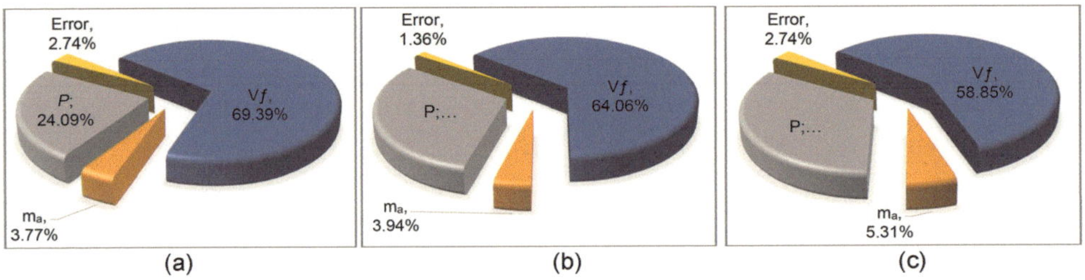

Figure 5.
Percentage contribution of variables for surface roughness. (a) R_a3 (μm) for t = 4 mm (b) R_a3 (μm) for t = 8 mm (c) R_a2 (μm) for t = 12 mm.

The impacts of the parameters for all profiles across the three levels of material thicknesses demonstrated a similar trend, denoting traverse speed and waterjet pressure to be significant factors for acquiring p-Values lower than 0.05, as detailed in **Tables A1-A3**. Accordingly, this work has established that abrasive mass flow rate is an insignificant input parameter for obtaining p-Values >0.05, ranging from 0.002 to 0.067. **Figure 5** represents the percentage contribution of variables for R_a of the most fitted regression models for 4, 8 and 12 mm material thickness. Overall, traverse speed features as the most influencing parameter, followed by waterjet pressure and abrasive mass flow rate. It can be observed here that the influence of traverse speed decreases, ranging from 69.39 to 58.85%, as the material thickness increases. In AWJM, an increasing traverse speed reduces the number of abrasive particles, leading to higher occurrences of surface roughness [47]. **Figure 5** shows that as the material thickness increases, the percentage contribution of waterjet pressure and abrasive mass flow rate also increases, ranging from 24.09 to 33.1% and 3.77 to 5.31%, respectively. The increasing value of waterjet pressure denotes higher energy, reinforcing a larger amount of abrasive particles obtaining lower surface roughness [48]. Further, an increasing rate of abrasive mass flow breaks down abrasive particles into smaller sizes, resulting in more sharp edges that reduce surface roughness [15]. The percentage errors obtained were less than 20%, indicating acceptable reliability of the models, as described in Eq. (6).

3.2 Regression model and analysis for material removal rate

Table 9 displays multi-linear regression coefficients of models developed for material removal rate against input parameters i.e., traverse speed (X_1), abrasive mass flow rate (X_2) and waterjet pressure (X_3) for 4, 8 and 12 mm material thicknesses of AISI 304L. Regardless of material thickness and cutting profile category, the input parameter coefficients acquired a positive sign whilst the constant coefficients had a negative sign. The sign of the coefficient denotes the trend of relationship between variables and response [44]. As a result, an increasing rate of traverse speed, abrasive mass flow rate and waterjet pressure, generates a higher rate of material removal. Overall, the coefficient of determination R^2 ranged from 97.79 to 97.92%, with R^2adj ranging from 96.46 to 96.67% and R^2pred ranging from 92.53 to 94.35%, confirming that all generated regression models were significant. The models were established to be sufficient for accurate forecasting of material removal rate within the assigned levels of input parameters for AWJM of straight and arcs profiles. Furthermore, **Table 9** demonstrated that MRR1 (straight-line), MRR2 (inner

Term	t = 4 mm			t = 8 mm			t = 12 mm		
	MRR1	MRR2	MRR3	MRR1	MRR2	MRR3	MRR1	MRR2	MRR3
	Coef	Coef	Coef	Coef	Coef	Coef	Coef	Coef	Coef
	−84.2	−33	−22.8	−119	−45	−60.6	−158.8	−43.3	−73.5
β_1	1.752	1.562	1.440	2.941	2.658	2.708	3.867	4.476	3.416
β_2	0.1260	0.0833	0.0901	0.2723	0.1738	0.0333	0.3960	0.205	0.2437
β_3	0.5103	0.3430	0.4101	0.7770	0.775	0.950	0.917	0.511	1.080
Model Summary									
R^2	97.92%	97.56%	97.54%	98.86%	97.71%	94.73%	98.70%	96.37%	97.79%
R^2 (adj)	96.67%	96.09%	96.06%	98.18%	96.33%	91.56%	97.92%	94.20%	96.46%
R^2 (pred)	94.35%	90.74%	91.12%	95.73%	91.90%	82.30%	95.19%	89.41%	92.53%

Table 9.
Summary of linear regression coefficients for MRR.

arcs) and MRR3 (outer arcs) attained a uniform gap of at least 2% for R^2, R^2adj and R^2pred values. This nominal disparity of the coefficient of determination indicates that AWJM performance for straight and curvature profiles are not significantly different from one another [36]. The results detailed in **Table 9** confirm that the highest values of R^2, R^2adj and R^2pred for all material thicknesses was attained in MRR1 (straight-line profile) with values of 97.92, 96.67 and 94.35%; 98.86, 98.18 and 95.73%; 98.70, 97.92 and 95.19% respectively. This statistical measurement evaluates the relationship between the model and response variables, indicating that a value nearest to 100% denotes a more reliable model [49]. Therefore, MRR1 regression models are considered as the most fitted model for 4, 8 and 12 mm material thicknesses.

Tables 10-12 present the predicted MRR values using the generated regression models of 4, 8 and 12 mm thickness of AISI 304L for three varied contour profiles. The percentage error acquired for 4, 8 and 12 mm AISI 304L thicknesses ranged from −5.35 to 5.15%, −6.59 to 4.77% and − 5.05 to 6.62%, respectively. The errors determined for R_a between the predicted value and experimental results were less than 20%, indicating models to be well fitted for predicting MRR values.

Plots of all residuals of the best material removal rate (MRR1) for all material thicknesses are represented in **Figure 6**. Overall, the normal probability plots for all the material thicknesses illustrate that the adjacency of the points are linear indicating there is no deviation from the assumptions, because they are normally and independently distributed [46]. Residuals versus fits and observation for MRR1 of straight-line, inner and outer arc profiles confirm that there is no skewness or outlier pattern, revealing that individual deviated assumptions have no conflicts or contradictions. **Figure 6** also presents the histogram graph for MRR1, obtaining frequency ranging from −10 to 15 for 4 mm, −15 to 15 for 8 mm and − 18 to 20 for 12 mm material thicknesses. These results signify that the distribution or frequency of residuals for all observations fell in minimal interval or inequalities of the experimental data, justifying the adequacy of the suggested MRR1 models [46].

According to the results presented in **Tables A4-A6** in the Appendix section, detailing ANOVA for material removal rate, the effects of the input parameters for

Exp. no.	Independent variables			MRR 1 (mm^3/min)			MRR 2 (mm^3/min)			MRR 3 (mm^3/min)		
	X_1	X_2	X_3	Exp.	Pred.	Error %	Exp.	Pred.	Error %	Exp.	Pred.	Error %
1	90	300	200	216.2	213.3	1.36	212.1	201.2	5.15	217.7	215.9	0.83
2	90	400	250	248.6	251.4	−1.10	223.1	226.7	−1.60	242.4	245.5	−1.27
3	90	500	300	284.2	289.5	−1.86	250.6	252.1	−0.62	267.8	275.0	−2.68
4	120	300	250	280.6	291.3	−3.82	251.7	265.2	−5.35	283.0	279.6	1.19
5	120	400	300	342.5	329.4	3.82	293.7	290.7	1.03	313.7	309.2	1.44
6	120	500	200	298.8	291.0	2.61	263.5	264.7	−0.45	286.2	277.2	3.14
7	150	300	300	372.1	369.4	0.73	333.8	329.2	1.38	343.9	343.4	0.16
8	150	400	200	330.7	330.9	−0.07	299.6	303.2	−1.21	298.5	311.4	−4.32
9	150	500	250	361.5	369.1	−2.09	333.5	328.7	1.44	344.8	340.9	1.14

Table 10.
Predicted MRR values of regression model for t = *4 mm.*

Exp. no.	Independent variables			MRR 1 (mm^3/min)			MRR 2 (mm^3/min)			MRR 3 (mm^3/min)		
	X_1	X_2	X_3	Exp.	Pred.	Error %	Exp.	Pred.	Error %	Exp.	Pred.	Error %
1	90	300	200	367.9	382.8	−4.05	405.0	401.4	0.88	399.0	383.0	4.00
2	90	400	250	456.9	448.9	1.75	450.8	457.6	−1.50	427.2	433.8	−1.56
3	90	500	300	511.2	515.0	−0.74	501.8	513.7	−2.37	493.1	484.7	1.71
4	120	300	250	526.9	509.9	3.23	526.4	519.9	1.23	488.1	511.7	−4.84
5	120	400	300	572.9	576.0	−0.54	583.5	576.1	1.27	579.8	562.6	2.97
6	120	500	200	532.9	525.5	1.39	532.2	515.9	3.06	441.8	470.9	−6.59
7	150	300	300	633.7	637.0	−0.52	639.7	638.4	0.19	629.1	640.5	−1.81
8	150	400	200	583.9	586.5	−0.45	555.1	578.3	−4.17	576.3	548.8	4.77
9	150	500	250	647.8	652.6	−0.74	641.3	634.5	1.07	601.3	599.6	0.28

Table 11.
Predicted MRR values of regression model for t = *8 mm.*

straight and arc profiles at 4, 8 and 12 mm AISI 304L thicknesses display comparable results. Further, the results reveal that traverse speed and waterjet pressure are statistically and physically significant factors for obtaining p-Values < 0.05. Hence, the abrasive mass flow rate features as a low impacting input parameter for obtaining p-Values greater than the acceptable value of 0.05, ranging from 0.002 to 0.751.

The percentage contribution of variables for the most fitted regression models MRR for 4, 8 and 12 mm material thicknesses are illustrated in **Figure 6**. In general, traverse speed is indicated as the most impacting variable, followed by waterjet pressure and abrasive mass flow rate, with a percent contribution ranging from 71.14–78.94%, 12.11–24.09% and 2.65–9.03% respectively for all profiles and material thicknesses. It is apparent here that the percentage contribution of traverse speed increases

Exp. no.	Independent variables			MRR 1 (mm^3/min)			MRR 2 (mm^3/min)			MRR 3 (mm^3/min)		
	X_1	X_2	X_3	Exp.	Pred.	Error %	Exp.	Pred.	Error %	Exp.	Pred.	Error %
1	90	300	200	472.0	491.5	−4.13	506.3	523.4	−3.37	528.9	523.1	1.10
2	90	400	250	586.2	576.9	1.58	542.1	569.5	−5.05	588.7	601.5	−2.17
3	90	500	300	655.9	662.4	−0.99	625.7	615.6	1.62	665.4	679.9	−2.18
4	120	300	250	676.0	653.4	3.35	731.7	683.2	6.62	687.5	679.6	1.15
5	120	400	300	735.0	738.8	−0.51	735.8	729.3	0.88	772.3	758.0	1.85
6	120	500	200	701.2	686.7	2.07	712.3	698.7	1.90	695.1	674.4	2.98
7	150	300	300	813.0	815.2	−0.27	822.4	843.1	−2.51	835.4	836.1	−0.09
8	150	400	200	755.6	763.1	−1.00	811.9	812.5	−0.07	725.0	752.5	−3.79
9	150	500	250	841.6	848.6	−0.83	845.6	858.6	−1.54	837.6	830.9	0.80

Table 12.
Predicted MRR values of regression model for t = *12 mm.*

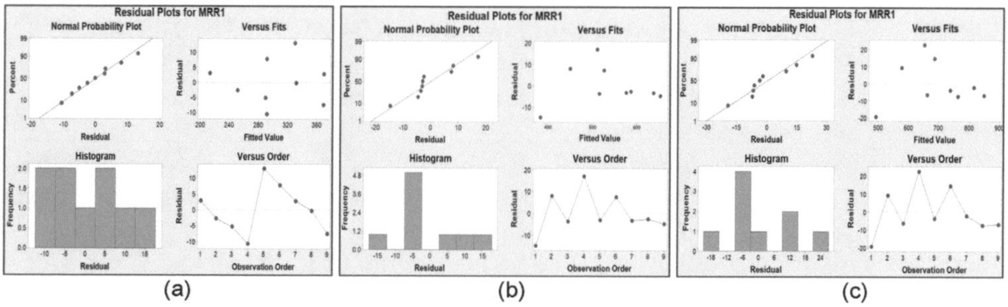

Figure 6.
Residual plots for material removal rate. (a) MRR 1 (mm^3/min) for t = *4 mm (b) MRR 1 (mm^3/min) for* t = *8 mm (c) MRR (mm^3/min) for* t = *12 mm.*

in range from 71.4 to 77.55% as the material thickness increases. An increasing traverse speed reinforces the contact time of the waterjet with the abrasive on the material, producing a higher volume rate of material to the machine [9]. Contrastingly, the percentage contribution of waterjet pressure and abrasive mass flow rate decreased as the material thickness and traverse speed increased, ranging from 22.42–12.11% and 4.35–9.03%, respectively. The increasing traverse speed and depth or thickness of the material to cut, results in a more prolonged machining process, which gradually leads to subsiding kinetic energy and loss of large of abrasive particles, resulting in reduced effectiveness of abrasive mass flow rate and waterjet pressure during the erosion process (**Figure 7**) [9, 47].

3.3 Regression model and analysis for kerf taper angle

The summary of the multi-linear regression coefficients for kerf taper angle of straight-line, inner and outer arc profiles using 4, 8 and 12 mm material thicknesses are detailed in **Table 13**. The results provide a similar trend, showing the constant sign

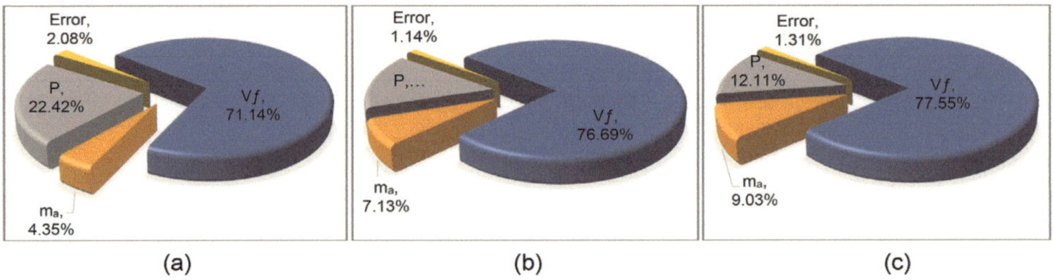

Figure 7.
Percentage contribution of variables for material removal rate. (a) MRR 1 (mm^3/min) for t = *4 mm (b) MRR 1 (mm^3/min) for* t = *8 mm (c) MRR (mm^3/min) for* t = *12 mm.*

as positive, with variables X_1, X_2 and X_3 as negative for all profiles and thicknesses. If the coefficient sign is negative, as the variable increases, the response decreases, whereas if the coefficient is positive, the relationship between variables and responses is directly proportional [44]. Therefore, an increasing rate of traverse speed (X1) results in an increasing angle of the kerf taper. Thus, an increasing rate of abrasive mass flow and waterjet pressure reduces the value of kerf taper angle. The values of R^2, R^2adj and R^2pred for 4, 8 and 12 mm ranged from 94.74–99.37%, 91.59–98.99% and 80.11–97.66%, respectively. This confirms that regression models are reliable in representing correlation between variables and responses and can be used in the optimisation of process parameters.

The coefficient of determination (R^2, R^2adj and R^2pred) obtained from straight-line, inner and outer arc profiles for all material thicknesses had a similar and consistent gap of at least 2%. The AWJM provides comparable behaviour in processing both straight and curvature profiles [36]. The highest values of R^2, R^2adj and R^2pred for 4 and 8 mm material thicknesses were attained in KTA1 with values of 97.56, 96.09 and 90.57%; 98.02, 96.82 and 92.01%; 99.37, 98.99 and 97.66%, respectively. These are the most fitted model, to be utilised in the optimisation of the process parameters of this study.

The predicted KTA values using the regression models applied for straight-line, inner and outer arc profiles of the three levels of material thicknesses are detailed in **Tables 14-16**. The percentage error obtained for 4, 8 and 12 mm AISI 304L thicknesses ranged between −2.55 to 1.72%, −2.67 to 3.74% and − 3.14 to 2.43%, respectively. The errors calculated for KTA between the predicted value and experimental results were less than the acceptable maximum limit of 20%, indicating the reliability of the models in predicting KTA values.

Figure 8 illustrates the residual plot for KTA including normal probability plot, residual versus fits, histogram for residuals and residuals versus experimental values. The results showed that the most fitted regression model is achieved from KTA1 for all material thicknesses. Correspondingly, the normal probability plots for all material thicknesses present a near fit to a line in a normal probability graph. The points constructing an approximate straight-line and plotted along the fitted line signifies that the data is normally distributed and there is a good relation between experimental data and predicted values [45]. Predominantly, the residuals versus fits and observation graph for each material thickness exhibit that the points are plotted randomly and near both sides of 0 with no identified pattern denoting a minimal deviation within residuals and estimated values. **Figure 8** also presents the histogram graph for KTA

Term	t = 4 mm			t = 8 mm			t = 12 mm		
	KTA1	KTA2	KTA3	KTA1	KTA2	KTA3	KTA1	KTA2	KTA3
	Coef	Coef	Coef	Coef	Coef	Coef	Coef	Coef	Coef
	0.9674	1.0469	1.064	1.386	1.483	1.544	1.5981	1.971	1.998
β_1	0.002414	0.002155	0.001501	0.006143	0.003594	0.004333	0.006568	0.004556	0.004736
β_2	- 0.000235	- 0.000220	- 0.000136	−0.00052	−0.000525	−0.00035	- 0.000107	- 0.000400	- 0.000436
β_3	- 0.000932	- 0.000952	- 0.000668	−0.002039	−0.001346	- 0.001867	- 0.002319	- 0.002320	- 0.002286
Model Summary									
R^2	97.56%	97.26%	94.74%	98.02%	94.76%	96.79%	99.37%	96.30%	96.95%
R^2 (adj)	96.09%	95.61%	91.59%	96.82%	91.61%	94.87%	98.99%	94.08%	95.12%
R^2 (pred)	90.57%	88.61%	84.48%	92.01%	80.11%	88.29%	97.66%	86.50%	88.70%

Table 13.
Summary of linear regression coefficients for KTA.

Exp. no.	Independent variables			KTA1 (°)			KTA2 (°)			KTA3 (°)		
	X_1	X_2	X_3	Exp.	Pred.	Error %	Exp.	Pred.	Error %	Exp.	Pred.	Error %
1	90	300	200	0.93	0.93	0.23	0.99	0.98	0.41	1.02	1.02	−0.19
2	90	400	250	0.86	0.86	0.26	0.92	0.92	0.84	0.98	0.98	0.15
3	90	500	300	0.77	0.79	−2.30	0.83	0.85	−2.16	0.94	0.93	0.53
4	120	300	250	0.95	0.95	−0.60	1.00	1.00	0.16	1.04	1.04	0.78
5	120	400	300	0.90	0.88	1.74	0.94	0.93	1.33	0.96	0.99	−2.55
6	120	500	200	0.97	0.95	1.72	1.00	1.01	−0.20	1.05	1.04	0.70
7	150	300	300	0.98	0.98	0.34	1.01	1.02	−0.80	1.06	1.05	1.13
8	150	400	200	1.03	1.05	−1.70	1.08	1.09	−1.44	1.09	1.10	−1.21
9	150	500	250	0.98	0.98	0.09	1.04	1.02	1.68	1.06	1.05	0.54

Table 14.
Predicted KTA values of regression model for t = *4 mm.*

Exp. no.	Independent variables			KTA1 (°)			KTA2 (°)			KTA3 (°)		
	X_1	X_2	X_3	Exp.	Pred.	Error %	Exp.	Pred.	Error %	Exp.	Pred.	Error %
1	90	300	200	1.38	1.38	0.04	1.40	1.38	1.42	1.43	1.46	−1.83
2	90	400	250	1.22	1.22	−0.12	1.27	1.26	0.76	1.34	1.33	0.91
3	90	500	300	1.04	1.07	−2.65	1.13	1.14	−0.94	1.21	1.20	0.87
4	120	300	250	1.48	1.46	1.50	1.43	1.42	0.43	1.54	1.49	3.07
5	120	400	300	1.35	1.30	3.42	1.30	1.30	−0.07	1.35	1.36	−1.07
6	120	500	200	1.44	1.46	−0.81	1.34	1.38	−3.20	1.49	1.52	−1.75
7	150	300	300	1.50	1.54	−2.67	1.44	1.46	−1.67	1.50	1.53	−1.96
8	150	400	200	1.68	1.69	−0.61	1.53	1.54	−0.81	1.70	1.68	1.11
9	150	500	250	1.56	1.54	1.41	1.48	1.42	3.74	1.56	1.55	0.46

Table 15.
Predicted KTA values of regression model for t = *8 mm.*

illustrating the distribution or frequency of the residuals for all observations. The results show that the frequency of KTA for 4, 8 and 12 mm material thicknesses range from −0.002 to 0.015, −0.05 to 0.05 for 8 mm and − 0.02 to 0.03, respectively. These graphs reveal a minimal interval or inequalities of the experimental data indicating that the KTA regression models are highly fitted to concrete prediction [46].

Tables A7-A9 in the Appendix section detail the results of ANOVA, where it can be observed that the impacts of parameters for all profiles and three levels of material thicknesses demonstrate a similar trend, denoting traverse speed and waterjet pressure to be significant factors for acquiring p-Values lower than 0.05. Thus, the abrasive mass flow rate was found insignificant for achieving p-Values >0.05, ranging from 0.002 to 0.245 for all profiles and material thicknesses.

Exp. no.	Independent variables			KTA1 (°)			KTA2 (°)			KTA3 (°)		
	X_1	X_2	X_3	Exp.	Pred.	Error %	Exp.	Pred.	Error %	Exp.	Pred.	Error %
1	90	300	200	1.70	1.69	0.39	1.80	1.80	0.18	1.85	1.84	0.49
2	90	400	250	1.57	1.57	0.07	1.63	1.64	−0.48	1.67	1.68	−0.26
3	90	500	300	1.43	1.44	−0.70	1.45	1.49	−2.20	1.49	1.52	−2.08
4	120	300	250	1.79	1.77	0.87	1.83	1.82	0.92	1.88	1.86	0.88
5	120	400	300	1.65	1.65	0.14	1.67	1.66	0.60	1.71	1.71	0.43
6	120	500	200	1.86	1.87	−0.72	1.90	1.85	2.43	1.92	1.89	1.51
7	150	300	300	1.84	1.86	−0.85	1.85	1.84	0.52	1.89	1.89	0.13
8	150	400	200	2.06	2.08	−0.81	1.97	2.03	−3.14	2.02	2.08	−2.93
9	150	500	250	1.98	1.95	1.51	1.89	1.87	0.82	1.95	1.92	1.55

Table 16.
Predicted KTA values of regression model for t = 12 mm.

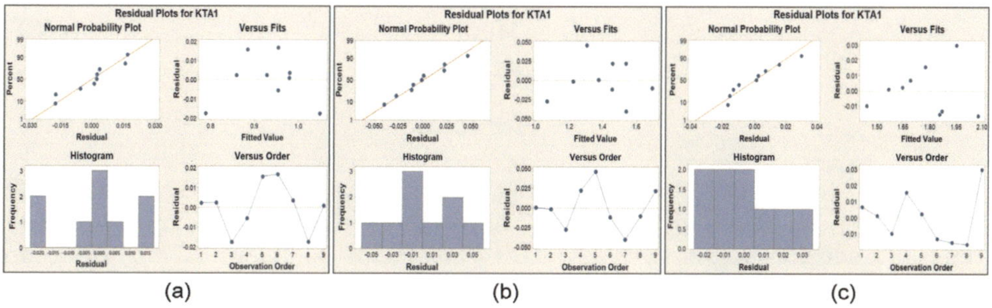

Figure 8.
Residual plots for kerf taper angle. (a) KTA1 (°) for t = 4 mm (b) KTA1 (°) for t = 8 mm (c) KTA1 (°) for t = 12 mm.

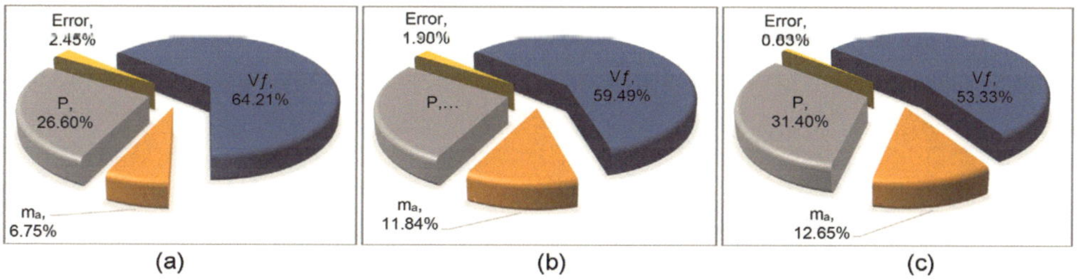

Figure 9.
Percentage contribution of variables for kerf taper angle. (a) t = 4 mm (b) t = 8 mm (c) t = 12 mm.

Figure 9 exhibits the percentage contribution of variables for KTA for the most fitted regression models for 4, 8 and 12 mm material thickness. Traverse speed was the most influencing parameter, followed by waterjet pressure and abrasive mass flow rate, in agreement with previous studies [14, 37]. The obtained results have shown that the influence of traverse speed decreases in range from 64.21 to 53.33% as the material thickness increases. An increasing value traverse speed results in the loss of a large number of abrasive particles, continuously dropping as the material thickness also increases, leading to a higher angle of kerf taper [50]. **Figure 9** shows increases of material thickness, the percentage contribution of waterjet pressure and abrasive mass flow rate, ranging from 26.60 to 33.40% and 6.75 to 12.65%, respectively. This increasing value of waterjet pressure resulted in higher energy, generating a larger amount of abrasive particles that result in a lower kerf taper [51]. Moreover, a rising rate of abrasive mass flow breaks down abrasive particles into a smaller scale, generating more sharp points that results in reduction of kerf taper angle [51].

4. Response surface methodology multi-objective optimisation

In this research, multi-objective optimisation was performed using RSM to determine the optimum process parameters of abrasive waterjet contour cutting of AISI 304L with varied thicknesses using MINITAB 19 software. The following optimisation objectives were stated as follows:

$$f_1 = \text{Min}\ (R_a) \tag{8}$$

$$f_2 = \text{Min}\ (\ KTA) \tag{9}$$

$$f_3 = \text{Max}\ (MRR) \tag{10}$$

RSM optimisation was performed using the models with the highest determination of coefficients, i.e., R^2, R^2adj and R^2pred. Accordingly, the regression models utilised to minimise surface roughness were R_a3 for 4 and 8 mm and R_a2 for 12 mm. MRR1 and KTA 1 models were used for all material thicknesses.

The Regression models utilised in multi-objective optimisation for varied thicknesses of AISI 304L were expressed by Eqs. (8)-(16).

$$\text{Ra}_{4\text{mm}} = 1.4256 + 0.003222\, X_1 - 0.000217\, X_2 - 0.001133\, X_3 \tag{11}$$

$$\text{KTA}_{4\text{mm}} = 0.9674 + 0.002414\, X_1 - 0.000235\, X_2 - 0.000932\, X_3 \tag{12}$$

$$\text{MRR}_{4\text{mm}} = -84.2 + 1.752\, X_1 + 0.126\, X_2 + 0.5103\, X_3 \tag{13}$$

$$\text{Ra}_{8\text{mm}} = 1.76 + 0.008869\, X_1 - 0.000577\, X_2 - 0.001920\, X_3 \tag{14}$$

$$\text{KTA}_{8\text{mm}} = 1.386 + 0.006143\, X_1 - 000520\, X_2 - 0.002039\, X_3 \tag{15}$$

$$\text{MRR}_{8\text{mm}} = -119 + 2.941 X_1 + 0.2723\, X_2 + 0.777 X_3 \tag{16}$$

$$\text{Ra}_{4\text{mm}} = 2.3854 + 0.004276\, X_1 - 0.000446\, X_2 - 0.001924\, X_3 \tag{17}$$

$$\text{KTA}_{4\text{mm}} = 1.5981 + 0.006568\, X_1 - 0.000107\, X_2 - 0.002319\, X_3 \tag{18}$$

$$\text{MRR}_{8\text{mm}} = -158.8 + 3.867 X_1 + 0.396\, X_2 + 0.917\, X_3 \tag{19}$$

In simultaneous optimisation, goals and boundaries must be defined for each process parameter. Targets are based on the experimental data obtained, referring to the set highest value of responses for maximising MRR and lowest value of responses for minimising R_a and KTA. In this optimisation, process parameters and defined objectives were assigned to be equally significant. Therefore, the equal weights (wt. = 1) were assigned in order to achieve an equal importance to the process parameters and objectives. The constraints referring to range and limits of the process parameters are detailed below.

Constraints:
$90 \leq V_f \leq 150$ mm/min
$300 \leq m_a \leq 500$ g/min
$200 \leq P \leq 300$ g/min

Limits:
$KTA_{4mm} \leq 1.03^{\circ}, KTA_{8mm} \leq 1.68^{\circ}, KTA_{12mm} \leq 2.06^{\circ}$

Ra_{4mm} mm1.58 μm,Ra_{8mm} mm2.45 μm,Ra_{12mm} 2m2.46 μm

$MRR_{4mm} \geq 216.20$ mm3/ min ,$MRR_{8mm} \geq 367.90$ mm3/ min ,
$MRR_{12mm} \geq 472.00$ mm3/ min

Table 17 shows the solutions for multi-objective optimisation performed for 4, 8 and 12 mm thickness of AISI 304L. The solution that provides the value of composite desirability nearest to 1 can be considered as the best solution [40]. **Table 17** reveals that solution 1 is the best for 4, 8 and 12 mm material thicknesses, achieving composite desirability values of 0.748448, 0.780587 and 0.786800, respectively. There are three solutions generated from MINITAB application, providing the settings of input variables, achieved values of responses and composite desirability. Solution 1 provides the optimum settings of input parameters i.e., V_f for 4, 8 and 12 mm material thicknesses, at the speeds of 95, 90 and 91 mm/min, respectively. The obtained optimum setting for m_a and P were found to be the same value for all material thicknesses, at 500 g/min and 200 MPa, respectively. **Table 17** presents the minimum achieved values of KTA and R_a and maximum MRR for 4, 8 and 12 mm material thicknesses, featuring at 0.799^0, 1.283 μm and 297.98 mm^3/min; 1.068^0, 1.694 μm and 514.97 mm^3/min and 1.448^0, 1.975 μm and 667.07 mm^3/min, respectively.

An optimisation plot presenting how the variables affected the predicted responses is shown in **Figure 10**, detailing the composite desirability for multi-objective (D) and single-objective optimisation (d). Current variable settings for the input parameters are presented in the figure, alongside with lower and upper limits. **Figure 10** shows a three-sectioned line graph representing the correlation of KTA, R_a and MRR against traverse speed (X_1), abrasive mass flow rate (X_2) and waterjet pressure (X_3).

From the figure, it can be observed that abrasive waterjet contour cutting responses demonstrate a comparable behaviour against input parameters for all material thicknesses. The highest rate of material removal and lowest value of surface roughness and Kerf taper angle were achieved by employing a rate of 150 mm/min speed, 500 g/min abrasive mass flow rate, and 300 MPa of waterjet pressure. Increasing water pressure, alongside high velocity abrasive mass flow rate, produces a greater collision of abrasive particles, generating higher rate of material removal and reducing surface roughness and kerf taper angle [52].

The surface roughness displayed an incrementing value that ranged from 4–13% as the rate of traverse speed increased from 90 to 150 mm/min. As the speed increases per unit of area over time, the kinetic energy containing abrasives gradually decreases, resulting in greater evidences of rough surfaces [52]. Consequently, RSM optimisation

Parameters	4 mm			8 mm			12 mm		
	Solutions								
	1	2	3	1	2	3	1	2	3
$X_1 = V_f$ (mm/min	95	97	97	90	90	116	91	90	90
$X_2 = m_a$ (g/min)	500	500	500	500	500	301.737	500	500	500
$X_3 = P$ (MPa)	300	300	300	300	300	300	300	300	300
KTA (°)	0.799	0.805	0.805	1.068	1.068	1.330	1.448	1.441	1.441
MRR (mm^3/min)	297.98	302.17	302.17	514.97	514.97	537.49	667.07	662.78	662.78
R_a (μm)	1.283	1.291	1.291	1.694	1.694	2.039	1.975	1.970	1.970
Composite Desirability	0.748448	0.748075	0.748075	0.780587	0.780587	0.556566	0.786800	0.786677	0.786677

Table 17.
Solutions for RSM multi-objective optimisation.

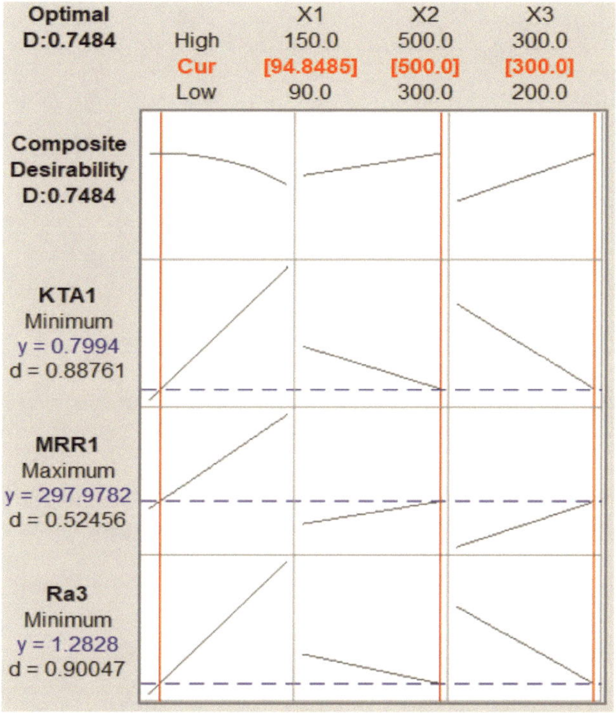
Optimal
D:0.7484
High
Cur
Low
X1
150.0
[94.8485]
90.0
X2
500.0
[500.0]
300.0
X3
300.0
[300.0]
200.0
Composite
Desirability
D:0.7484
KTA1
Minimum
y = 0.7994
d = 0.88761
MRR1
Maximum
y = 297.9782
d = 0.52456
Ra3
Minimum
y = 1.2828
d = 0.90047

(a)

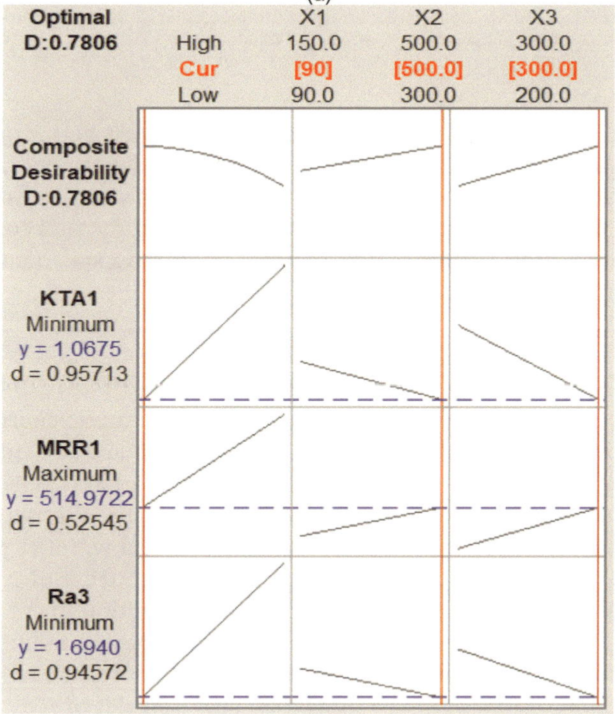
Optimal
D:0.7806
High
Cur
Low
X1
150.0
[90]
90.0
X2
500.0
[500.0]
300.0
X3
300.0
[300.0]
200.0
Composite
Desirability
D:0.7806
KTA1
Minimum
y = 1.0675
d = 0.95713
MRR1
Maximum
y = 514.9722
d = 0.52545
Ra3
Minimum
y = 1.6940
d = 0.94572

(b)

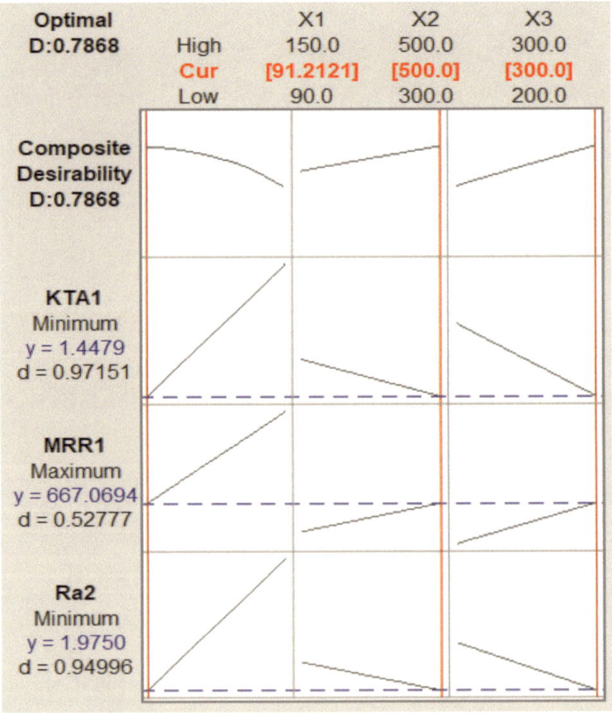

(c)

Figure 10.
Response optimisation plot. (a) t = 4 mm (b) t = 8 mm (c) t = 12 mm.

has shown that a lower level of traverse speed can produce a better quality of cut surface. Additionally, surface roughness in this study shows an increasing value ranging 2–5%, as the waterjet pressure increases and the abrasive mass flow rate from 200 to 300 MPa and 300 to 500 g/min, respectively. In this study, it is confirmed that augmenting abrasive flow rate and waterjet pressure, up to a specific range, lowers the value of surface roughness. When higher values of traverse speed are employed, the material removal exhibits an increasing rate that ranges from 16–20%. In addition, increasing rate of material removal was achieved with a range of 5–9%, as the rate of abrasive mass flow and waterjet pressure increased from 200 to 300 MPa and 300 to 500 g/min, respectively. AWJM produces a high level of kinetic energy, driving a higher level of speed and waterjet pressure alongside with abrasive mass flow rate, which in turn generates higher cutting area per unit of time and generates a larger amount of eroded material [53]. Therefore, the rate of material removal is directly proportional to traverse speed, abrasive mass flow rate and waterjet pressure.
Figure 10 shows that kerf taper angle values increase as the rate of traverse speed increases from 90 to 150 mm/min. With continuous reduction in the number of abrasive particles, as the traverse speed increases, the cohesion on metal material decreases, generating a higher tapering angle [52]. The kerf taper angle in this study was reduced by 2–7%, as the abrasive mass flow and waterjet pressure were increased from 200 to 300 MPa and 300 to 500 g/min, respectively. A higher waterjet pressure alongside with abrasive mass flow rate reinforces the collision of abrasive particles on the target material, causing the reduction of kerf taper angle [51].

5. Conclusions

This study focuses on modelling and establishing optimum abrasive waterjet contour cutting parameters that lead to minimum surface roughness, kerf taper angle and maximum productivity (material removal rate). On the basis of the results achieved and discussed, the following conclusions are accomplished:

1. The experimental results indicate that abrasive waterjet contour cutting responses demonstrate similar behaviour against input parameters for straight-line and curvature profiles. The correlation coefficients of the predictive models of R^2, R^2adj and R^2pred for surface roughness, kerf taper angle and material removal rate were found to be in the range of 88.66–99.08%, 82.3–98.86% and 82.3–98.86% respectively. Therefore, the developed multi-linear regression models are reliable and effective for predicting output responses, where the percentage errors are at minimum values ranging from −6.59 to 6.71%

2. The results of the ANOVA for R_a. MRR and KTA demonstrate that traverse speed is the most influencing factor, with percentage contributions ranging from 55.67 to 78.94%. Surface roughness and kerf taper angle decrease as waterjet pressure and abrasive mass flow rate increase, resulting in reductions ranging from 2–5% and 2–7%, respectively. Increasing values of traverse speed, waterjet pressure and abrasive mass flow rate lead to increased rates of material removal, ranging from 16–20% and 5–9%, respectively.

3. The multi-objective optimization was performed using RSM for optimising abrasive waterjet contour cutting process parameters applied for 4, 8 and 12 mm material thicknesses, achieving the highest composite desirability values of 0.748448, 0.780587 and 0.786800, respectively. The optimum settings of input parameters i.e., V_f for 4, 8 and 12 mm material thickness are 95, 90 and 91 mm/min, respectively. The obtained optimum settings for m_a and P were found to be the same value for all material thicknesses, at 500 g/min and 200 MPa, respectively. The minimum achieved values of KTA and R_a and maximum MRR for 4, 8 and 12 mm material thickness were 0.799^0, 1.283 μm and 297.98 mm^3/min; 1.068^0, 1.694 μm and 514.97 mm^3/min; and 1.448^0, 1.975 μm and 667.07 mm^3/min, respectively.

Abbreviations and nomenclature

ht	depth of cut (mm)
m_a	abrasive mass flow rate (g/min)
P	water pressure (MPa)
R_a	surface roughness (μm)
V_f	traverse speed (mm/min)
W	kerf width (mm)
W_t	kerf top width (mm)
W_b	kerf bottom width (mm)
t	thickness of the material (mm)
AISI	austenitic stainless steel

ANOVA analysis of variance
AWJM abrasive waterjet machining
KTA kerf taper angle (^{0})
MRR material removal rate (mm^3/min)

A. Appendix

Source	R_a 1		R_a 2		R_a 3	
	Contribution %	***p*-Value**	**Contribution %**	***p*-Value**	**Contribution %**	***p*-Value**
X_1	59.90	0.001	69.43	0.000	69.39	0.000
X_2	5.16	0.067	3.49	0.068	3.77	0.017
X_3	30.19	0.002	23.86	0.002	24.09	0.001
Error	4.74		3.23		2.74	
Total	100.00		100		100	

Table A1.
ANOVA of R_a for t = *4 mm.*

Source	R_a 1		R_a 2		R_a 3	
	Contribution %	***p*-Value**	**Contribution %**	***p*-Value**	**Contribution %**	***p*-Value**
X_1	72.21	0.000	71.07	0.000	64.06	0.000
X_2	7.96	0.007	11.57	0.003	3.94	0.013
X_3	17.84	0.001	15.52	0.001	30.64	0.001
Error	1.99		1.84		1.36	
Total	100.00		100.00		100.00	

Table A2.
ANOVA of R_a for t = *8 mm.*

Source	R_a 1		R_a 2		R_a 3	
	Contribution %	***p*-Value**	**Contribution %**	***p*-Value**	**Contribution %**	***p*-Value**
X_1	57.23	0.000	58.85	0.000	57.21	0.001
X_2	3.44	0.026	5.31	0.002	17.66	0.011
X_3	34.59	0.000	33.1	0.000	19.47	0.009
Error	4.74		2.74		3.23	3.23
Total	100.00		100.00		100.00	

Table A3.
ANOVA of R_a for t = *12 mm.*

Source	MRR 1		MRR 2		MRR 3	
	Contribution %	*p*-Value	Contribution %	*p*-Value	Contribution %	*p*-Value
X_1	71.14	0.000	70.98	0	75.503	0
X_2	4.35	0.023	2.65	0.067	3.345	0.048
X_3	22.42	0.002	23.93	0.001	18.688	0.002
Error	2.08		2.44		2.464	
Total	100.00		100.00		100.00	

Table A4.
ANOVA of MRR for t = *4 mm.*

Source	MRR 1		MRR 2		MRR 3	
	Contribution %	*p*-Value	Contribution %	*p*-Value	Contribution %	*p*-Value
X_1	76.69	0.000	76.12	0.000	70.51	0.000
X_2	7.13	0.002	3.62	0.038	0.12	0.751
X_3	15.05	0.000	17.98	0.002	24.09	0.005
Error	1.14		2.29		5.27	
Total	100.00		100.00		100.00	

Table A5.
ANOVA of MRR for t = *8 mm.*

Source	MRR 1		MRR 2		MRR 3	
	Contribution %	*p*-Value	Contribution %	*p*-Value	Contribution %	*p*-Value
X_1	77.55	0.000	78.94	0.000	73.29	0.000
X_2	9.03	0.002	4.13	0.04	4.15	0.028
X_3	12.11	0.001	13.3	0.001	20.35	0.001
Error	1.31		3.63		2.21	
Total	100.00		100.00		100.00	

Table A6.
ANOVA of MRR for t = *12 mm.*

Source	KTA 1		KTA 2		KTA 3	
	Contribution %	*p*-Value	Contribution %	*p*-Value	Contribution %	*p*-Value
X_1	64.21	0.000	58.7	0.000	56.74	0.000
X_2	6.75	0.014	6.77	0.017	5.27	0.075
X_3	26.6	0.001	31.78	0.001	32.74	0.001
Error	2.45		2.74		5.26	
Total	100.00		100.00		100.00	

Table A7.
ANOVA of KTA for t = *4 mm.*

Source	KTA 1		KTA 2		KTA 3	
	Contribution %	p-Value	Contribution %	p-Value	Contribution %	p-Value
X_1	59.49	0.001	67.49	0.000	60.95	0.000
X_2	11.84	0.015	5.63	0.013	4.42	0.047
X_3	26.69	0.002	21.65	0.001	31.42	0.001
Error	1.98		5.24		3.21	
Total	100.00		100.00		100.00	

Table A8.
ANOVA of KTA for t = *8 mm.*

Source	KTA 1		KTA 2		KTA 3	
	Contribution %	p-Value	Contribution %	p-Value	Contribution %	p-Value
X_1	53.33	0.000	70.59	0.000	55.67	0.000
X_2	12.65	0.055	0.22	0.245	5.24	0.033
X_3	31.40	0.001	25.50	0.000	36.04	0.001
Error	0.63		3.70		3.05	
Total	100.00		100.00		100.00	

Table A9.
ANOVA of KTA for t = 12 mm.

Author details

Jennifer Milaor Llanto, Ana Vafadar and Majid Tolouei-Rad*
School of Engineering, Edith Cowan University, Joondalup, Australia

*Address all correspondence to: m.rad@ecu.edu.au

References

[1] Radovanović M. Multi-objective optimization of process performances when cutting carbon steel with abrasive water jet. Tribology in Industry. 2016;**38**(4):1741-1759

[2] Liu X et al. Waterjet machining and research developments: A review. The International Journal of Advanced Manufacturing Technology. 2019; **102**(5):1257-1335

[3] Rajurkar K et al. Review of sustainability issues in non-traditional machining processes. Procedia Manufacturing. 2017;**7**:714-720

[4] Natarajan Y et al. Abrasive water jet machining process: A state of art of review. Journal of Manufacturing Processes. 2020;**49**:271-322

[5] Singh P et al. Developments of non-conventional drilling methods—a review. The International Journal of Advanced Manufacturing Technology. 2020;**106**(5):2133-2166

[6] Sureban R, Kulkarni VN, Gaitonde V. Modern optimization techniques for advanced machining processes—a review. Materials Today: Proceedings. 2019;**18**:3034-3042

[7] Rao RV, Rai DP, Balic J. Optimization of abrasive waterjet machining process using multi-objective jaya algorithm. Materials Today-Proceedings. 2018;**5**(2): 4930-4938

[8] Nair A, Kumanan S. Multi-performance optimization of abrasive water jet machining of Inconel 617 using WPCA. Materials and Manufacturing Processes. 2017;**32**(6):693-699

[9] Chakraborty S, Mitra A. Parametric optimization of abrasive water-jet machining processes using grey wolf optimizer. Materials and Manufacturing Processes. 2018;**33**(13):1471-1482

[10] Trivedi P, Dhanawade A, Kumar S. An experimental investigation on cutting performance of abrasive water jet machining of austenite steel (AISI 316L). Advances in Materials and Processing Technologies. 2015;**1**(3–4):263-274

[11] Gnanavelbabu A et al. Experimental investigations on multiple responses in abrasive waterjet machining of Ti-6Al-4V alloy. Materials Today: Proceedings. 2018;**5**(5):13413-13421

[12] Jeykrishnan J et al. Optimization of process parameters in abrasive water jet machining/cutting (AWJM) of nickel alloy using traditional analysis to minimize kerf taper angle. Materials Today: Proceedings. 2019;**16**:392-397

[13] Singh D, Shukla RS. Investigation of kerf characteristics in abrasive water jet machining of inconel 600 using response surface methodology. Defence Science Journal. 2020;**70**(3):313-322

[14] Madankar A et al. Estimation and control of surface quality and traverse speed in abrasive water jet machining of AISI 1030 steel using different work-piece thicknesses by RSM. Australian Journal of Mechanical Engineering. 2021: **1**:1-8

[15] Llanto JM et al. Recent progress trend on abrasive waterjet cutting of metallic materials: A review. Applied Sciences. 2021;**11**(8):3344

[16] Pérez L, Carmelo J. On the application of a design of experiments along with an anfis and a desirability function to model response variables. Symmetry. 2021;**13**(5):897

[17] Luis Pérez C. A proposal of an adaptive neuro-fuzzy inference system for modeling experimental data in manufacturing engineering. Mathematics. 2020;**8**(9):1390

[18] Jiang W et al. Effects of nanostructural hierarchy on the hardness and thermal stability of an austenitic stainless steel. Journal of Materials Research and Technology. 2021;**12**: 376-384

[19] Ramana MV et al. Optimization and influence of process parameters of dissimilar SS304L–SS430 joints produced by Robotic TIG welding. Materials Today: Proceedings. 2020;**23**: 479-482

[20] Doreswamy D et al. Machining of d2 heat treated steel using abrasive water jet: The effect of standoff distance and feed rate on kerf width and surface roughness. International Journal of Research in Engineering and Technology. 2014;**3**(8):417-421

[21] Shibin R et al. Investigation on the abrasive water jet machinability of AA2014 using SiC as abrasive. Materials Today: Proceedings. 2020;**21**:519-522

[22] Radovanovic M, Herghelegiu E. Perpendicularity deviation and surface roughness in abrasive water jet cutting of carbon steel. Revista de Tehnologii Neconventionale. 2016;**20**(2):39

[23] Maneiah D et al. Optimization of machining parameters for surface roughness during abrasive water jet machining of aluminium/magnesium hybrid metal matrix composites. Materials Today: Proceedings. 2020;**27**: 1293-1298

[24] Rajamanickam S et al. Comparative analysis of MRR on abrasive water jet machining parameters over aerospace alloys: Inconel 825 & Ti-6Al-4V. International Journal of Pure and Applied Mathematics. 2018;**118**:727-733

[25] Kaladhar M, Subbaiah KV, Rao CS. Machining of austenitic stainless steels–a review. International Journal of Machining and Machinability of Materials. 2012;**12**(1):178-192

[26] Hlaváč LM et al. Investigation of the taper of kerfs cut in steels by AWJ. The International Journal of Advanced Manufacturing Technology. 2015; **77**(9–12):1811-1818

[27] Pawar PJ, Vidhate US, Khalkar MY. Improving the quality characteristics of abrasive water jet machining of marble material using multi-objective artificial bee colony algorithm. Journal of Computational Design and Engineering. 2018;**5**(3):319-328

[28] Hlaváč LM et al. Shape distortion reduction method for abrasive water jet (AWJ) cutting. Precision Engineering. 2018;**53**:194-202

[29] Aamir M et al. Optimization and modeling of process parameters in multi-hole simultaneous drilling using taguchi method and fuzzy logic approach. Materials. 2020;**13**(3):680

[30] Msomi V, Mabuwa S. Analysis of material positioning towards microstructure of the friction stir processed AA1050/AA6082 dissimilar joint. Advances in Industrial and Manufacturing Engineering. 2020;**1**: 100002

[31] Lin C. Use of the Taguchi method and grey relational analysis to optimize turning operations with multiple performance characteristics. Materials and Manufacturing Processes. 2004; **19**(2):209-220

[32] Aamir M et al. Feasibility of tool configuration and the effect of tool material, and tool geometry in multi-hole simultaneous drilling of Al2024. The International Journal of Advanced Manufacturing Technology. 2020;**111**(3):861-879

[33] Nagaraj Y et al. Prediction of material removal rate and surface roughness in hot air assisted hybrid machining on soda-lime-silica glass using regression analysis and artificial neural network. Silicon. 2020:**13**(11):1-13

[34] Aydin G, Karakurt I, Hamzacebi C. Artificial neural network and regression models for performance prediction of abrasive waterjet in rock cutting. The International Journal of Advanced Manufacturing Technology. 2014;**75**(9–12):1321-1330

[35] Cetin MH et al. Evaluation of vegetable based cutting fluids with extreme pressure and cutting parameters in turning of AISI 304L by Taguchi method. Journal of Cleaner Production. 2011;**19**(17–18):2049-2056

[36] Llanto JM et al. Analysis and optimization of process parameters in abrasive waterjet contour cutting of AISI 304L. Metals. 2021;**11**(9):1362

[37] Llanto JM et al. Impacts of traverse speed and material thickness on abrasive waterjet contour cutting of austenitic stainless steel AISI 304L. Applied Sciences. 2021;**11**(11):4925

[38] Aamir M et al. Machinability of Al2024, Al6061, and Al5083 alloys using multi-hole simultaneous drilling approach. Journal of Materials Research and Technology. 2020;**9**(5):10991-11002

[39] Koli Y, Yuvaraj N, Aravindan S. Multi-response mathematical model for optimization of process parameters in CMT welding of dissimilar thickness AA6061-T6 and AA6082-T6 alloys using RSM-GRA coupled with PCA. Advances in Industrial and Manufacturing Engineering. 2021;**2**:100050

[40] Kumar KR, Sreebalaji V, Pridhar T. Characterization and optimization of abrasive water jet machining parameters of aluminium/tungsten carbide composites. Measurement. 2018;**117**:57-66

[41] Chabbi A et al. Predictive modeling and multi-response optimization of technological parameters in turning of polyoxymethylene polymer (POM C) using RSM and desirability function. Measurement. 2017;**95**:99-115

[42] Chate GR et al. Study of the effect of nano-silica particles on resin-bonded moulding sand properties and quality of casting. Silicon. 2018;**10**(5):1921-1936

[43] Javed SA et al. A critical review: Shape optimization of welded plate heat exchangers based on grey correlation theory. Applied Thermal Engineering. 2018;**144**:593-599

[44] Ratner B. The correlation coefficient: Its values range between+ 1/ − 1, or do they? Journal of Targeting, Measurement and Analysis for Marketing. 2009;**17**(2):139-142

[45] Deshpande Y, Andhare A, Sahu NK. Estimation of surface roughness using cutting parameters, force, sound, and vibration in turning of Inconel 718. Journal of the Brazilian Society of Mechanical Sciences and Engineering. 2017;**39**(12):5087-5096

[46] Galpin JS, Hawkins DM. The use of recursive residuals in checking model fit in linear regression. The American Statistician. 1984;**38**(2):94-105

[47] Sasikumar K et al. A study on kerf characteristics of hybrid aluminium 7075 metal matrix composites machined using abrasive water jet machining technology. Proceedings of the Institution of Mechanical Engineers, Part B: Journal of Engineering Manufacture. 2018;**232**(4): 690-704

[48] Kechagias J, Petropoulos G, Vaxevanidis N. Application of Taguchi design for quality characterization of abrasive water jet machining of TRIP sheet steels. The International Journal of Advanced Manufacturing Technology. 2012;**62**(5–8):635-643

[49] Andrzej P. Experimental research into alternative abrasive material for the abrasive water-jet cutting of titanium. The International Journal of Advanced Manufacturing Technology. 2018;**97** (1–4):1529-1540

[50] Uthayakumar M et al. Machinability of nickel-based superalloy by abrasive water jet machining. Materials and Manufacturing Processes. 2016;**31**(13): 1733-1739

[51] Babu MN, Muthukrishnan N. Exploration on Kerf-angle and surface roughness in abrasive waterjet machining using response surface method. Journal of The Institution of Engineers (India): Series C. 2018;**99**(6): 645-656

[52] Ishfaq K et al. Abrasive waterjet cutting of cladded material: Kerf taper and MRR analysis. Materials and Manufacturing Processes. 2019;**34**(5): 544-553

[53] Kmec J et al. The predictive model of surface texture generated by abrasive water jet for austenitic steels. Applied Sciences. 2020;**10**(9):3159

Chapter 3

Inkjet Printing of Functional Inks for Smart Products

Cláudia Buga and Júlio C. Viana

Abstract

Inkjet printing is a recent promising technology for direct patterning of solution-based materials over different substrates. It is particularly interesting for applications in the flexible electronics field and smart products manufacturing, as it allows for rapid prototyping, design freedom, and is compatible with conductive, semiconductive, and dielectric inks that can be cured at low temperatures over several types of substrates. Moreover, the inkjet process allows for ink economization, since great electrical conductivity can be achieved despite the deposition of small volumes of ink. This chapter describes the overall process, the main inks and their features, the critical process variables, and its limitations. Applications related to inkjet printing of functional materials and smart products are highlighted. New technology advancements and trends are finally addressed.

Keywords: inkjet printing, functional inks, process parameters, printing process optimization, inkjet printed devices, smart products

1. Introduction

Inkjet printing (IJP) is a widespread technology used in personal and industrial printers. Recently, it has started to gain traction as a new promising technology for the direct patterning of solution-based functional materials. IJP relies on a non-impact dot-matrix printing technology in which droplets of ink are flown from small openings, called nozzles, directly to a designated position on a media to produce an image. The printed patterns are digitally defined and directly transferred to the printer. The nozzles or the substrate holder move accordingly to a pre-programmed pattern, which allows the printing of virtually any pattern [1, 2].

IJP technique is particularly interesting for applications in the printed electronics field as it allows for rapid prototyping and is compatible with various substrates, and conductive, semiconductive, and dielectric inks that can be cured at low temperatures. As a result, several application examples of this technology have already been advanced in the literature [3–9]. IJP methods are widely employed in the manufacture of sensors and actuators, and many electrically conductive inks are already commercially available and optimized according to specific characteristics that make them suitable for IJP [10]. Since IJP relies on the use of computer software, it allows for rapid prototyping and freedom of design combined with tunable resolution [11]. Throughout the literature, some examples include IJP temperature sensors [12–14], humidity sensors [15–17], and pressure and strain sensors, that can be capacitive [18, 19], piezoresistive [20, 21], and piezoelectric [22–24]. These sensors and actuators can be integrated into novel smart products.

According to recent technical reports, printed electronics (PE), particularly IJP and its enabling technologies (functional printing and inks) show an increased market interest and growth. **Table 1** presents the market expectations related to these technologies.

PE market will continue to expand over the coming years, with a strong emphasis on energy harvesting and storage for electronic cars, gadgets, equipment, components, and other industries that rely on PE to reduce total energy usage. The PE market will be driven by low production costs, environmentally friendly technologies, a diverse choice of substrates, and a rising demand for flexible electronics applications. PE has allowed printing of electronic and electrical components on lightweight, cost-efficient and flexible materials (like cloth, paper, or polymeric films) in conventional electrical circuits. This PE market will continue to expand mainly driven by [29]: (a) the increasing development of smart and connected devices, as demanded by Internet of Things (IoT); (b) the rising demand for energy-effective, thin, and flexible consumer electronics; (c) the substantial costs reduction provided by PE; and (d) the importance of environmentally sustainable technologies.

The increased demand for low-cost and high-volume production of electronics will boost the functional printing market. This is supported by the increased availability of a wide range of substrates, high-throughput manufacturing technologies (e.g., R2R for large-area electronics processing), and a reduced environmental impact (e.g., thin and flexible electronics) [30]. This will be fostered by the development of new products/applications, the introduction of added functionalities into multiple products, and the emergent widespread of digital manufacturing techniques.

After a huge pace of growth, the IJP market is becoming mature, and high-speed inkjet printing devices with enhanced quality and higher productivity are already available. Notwithstanding, the initial costs of equipment are still rather high. Major applications of IJP have been related to graphic communication and packaging labeling, but functional substrates and objects driven by PE and functional printing applications are fostering IJP market. Principal drivers for the adoption of IJP technologies

	Printed electronics [25]	Functional printing [26]	Inkjet Printing [27]	Functional inks [28]
2020 market value, B$	7.9	8.9	40.8	0.87
2027 market value, B$	22.7	23.9	49.2	1.3
CAGR, %	21.5	15.1	3.11	5.3
Typical applications	Batteries, sensors, sign boards, labels, PCB, touch panels, LED panels, solar cells	Sensors, displays, batteries, RFID tags, lighting, photovoltaic, electronic components	Flexible OLED displays, wearables, photovoltaic, sensors, PCB	PCB, MEMS, security printing, smart textiles, displays, smart packaging, RFID tags, photovoltaics, biochips

Table 1.
Market values and growth rates for different related technologies: Printed electronics, functional printing, inkjet printing, functional inks, and smart products.

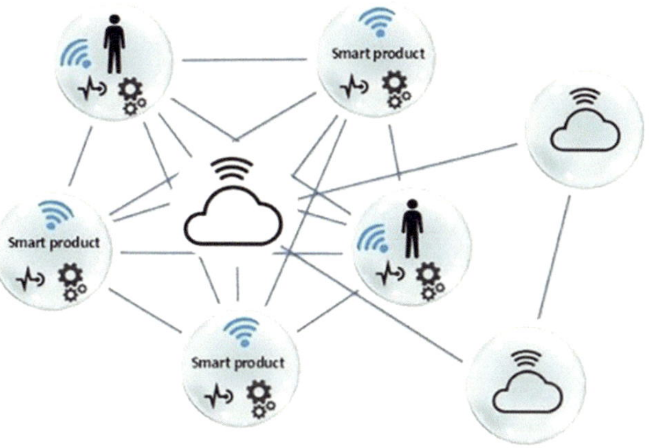

Figure 1.
System of smart products.

are OLED displays, products/processes digitalization, IoT, Cyber-physical systems (CPS), and Big Data.

Functional inks can be electrically conductive, resistive, dielectric, semi-conductive, or have other special functions, such as thermal conductivity, electroluminescence, light-diffusing, or piezoelectric. Functional inks are key enablers for PE applications. They must combine their functionality with being flexible, processable at low temperature, adhere to a wide range of substrates, in some cases transparent, and straightforward to manufacture. Novel functional inks include suspensions of organic or inorganic nanomaterials, or particle-free solutions of organic materials, which are inherently stretchable, and suitable for applications in e-textiles and in-mold electronics [31].

Smart products are physical objects equipped with sensors, embedded artificial intelligence, communication ability, and information technology. They bridge the physical and digital worlds, sharing information about themselves, their environment, and their use, being supported by emergent technologies of CPS, IoT, and artificial intelligence (AI). Furthermore, smart products are now connected and able of forming product eco-systems; they interact with the user, adding a social layer to these eco-systems. This allows a paradigm shift in the business world: from selling products to offering services, to the "servitization" of products (**Figure 1**). This transformation towards novel smart products is enabled by the development of emergent technologies simultaneously in both the physical (hardware) and digital (software) worlds, and their interfaces. This chapter focuses attention on product manufacturing (hardware) technologies for smart products, namely those based on PE and functional printing, and more specifically on inkjet printing of functional inks.

2. Inkjet printing of functional inks

2.1 Inkjet printing process

IPJ is an additive manufacturing technique that encompasses an ink reservoir that is connected to a print head device and responsible for jetting ink droplets over a pre-determined substrate. IJP allows for high-resolution 2D patterning, ink

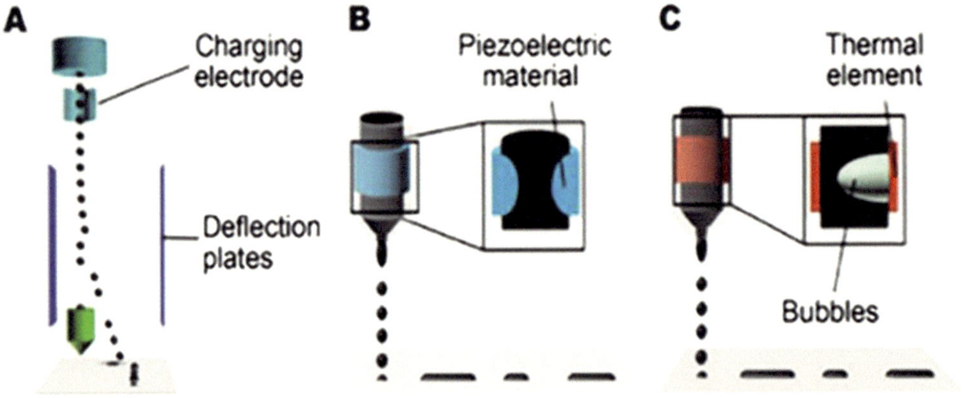

Figure 2.
IJP methods: (a) CIJ, and DOD inkjet printing with (b) piezoelectric and (c) thermal head [34].

economization, and non-contact deposition via a micrometer-sized inkjet nozzle head [32, 33]. Inkjet can be divided into two main distinct processes (**Figure 2**): Continuous inkjet (CIJ) printing and Drop-on-demand (DoD) printing technologies [35]. As the name suggests, in CIJ printing the droplets are continuously generated and deposited when subjected to an electrostatic field, caused by a charging electrode. DoD printing, on the other hand, relies on the selective activation of the print-head through impulses that can be acoustic, electrostatic, thermal, and piezoelectric (the latter two are the most reported cases) [35].

The CIJ process is mainly used in industrial printers, mostly for packaging and graphical applications. In this case, the ink droplets are continuously expelled due to the effect of an electric field that actuates the piezoelectric crystal of the print-head. Although this process can be used for PE it is majorly directed at continuously printing large volumes of nonfunctional inks [36]. Concerning DoD inkjet printing, several sub-methods can be identified. The most disseminated ones are piezoelectric and thermal inkjet printing, nonetheless, there are other methods whose popularity is increasing and can also grant high-quality printing of functional inks [36]. Among those, electrohydrodynamic (EHD)-IJP [37–39], aerosol jet printing [40], drop impact printing [41], and acoustic printing [42], can be highlighted.

In the piezoelectric IJP method, the ink reservoir is coupled with piezoelectric constrictors that load and expel the ink (print head). In this process, the dimensions of the ink droplet can be controlled, so the ink consumption is very low. To avoid clogging the nozzles, the functional inks must be produced taking into account specific properties, such as particle size, ink viscosity, surface tension, and density [35]. The nozzle is designed to be resistant to organic solvents and is therefore compatible with a wide range of solvents for ink formulation [2]. **Table 2** summarizes the main characteristics of the inks that are compatible with each one of the inkjet printing methods and sub-methods. Comparing the different printing sub-methods, it is clear that the thermal and the piezoelectric ones are much more limited in terms of suitable ink viscosity range. Nonetheless, they are still currently the most approachable methods in terms of affordability and widespread commercial availability of the equipment.

Except for the jetting method, the overall process of printing is common to all CIJ and DoD IJP techniques. The printed patterns are digitally generated (CAD software)

Printing Method	Sub-method	Particle size (nm)	Viscosity (cP)	Surface Tension (dynes/cm)	Resolution (μm)
CIJ	—	<1000	1–10	25–70	
DoD	Thermal	<1000	5–30	35–70	2–100
	Piezoelectric	<300	1–30	35–70	2–100
	EHD	3–300	1–4000	35–70	0.2–1
	Aerosol	10–30	0.5–2500	20–70	5–20
	Acoustic	<100	0.5–25,000	15–650	10–20
	Drop impact	<20,000	0.5–33	32–70	40–960

Table 2.
CIJ and DoD printing characteristics [2, 35, 37–43].

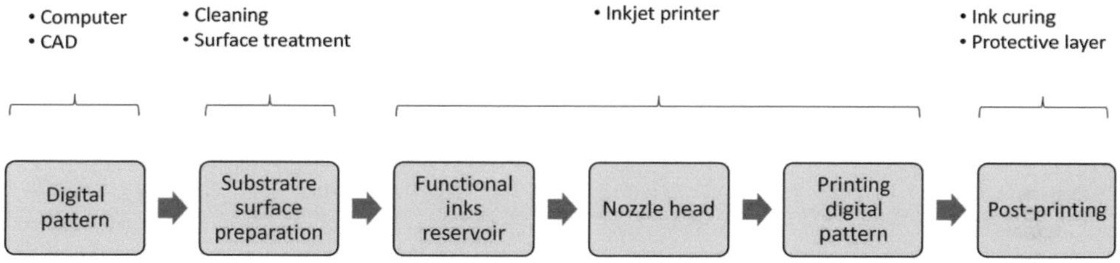

Figure 3.
Inkjet printing steps.

and can be easily changed, which makes this printing technique an ideal choice for prototyping and design optimization. **Figure 3** depicts the main stages in IJP:

1. Digital pattern—A CAD software (or other graphic software) is used to digitally generate the printed pattern.

2. Substrate surface preparation—the substrate surface is cleaned with an appropriate cleaning agent; a surface treatment can be selected for better ink adhesion.

3. Ink preparation—ink can be filtered to remove impurities; ink viscosity can be changed by using an adequate solvent; a surface treatment technique can be applied (e.g., plasma, corona discharge); the ink must be stirred for better particle suspension dispersion.

4. Inkjet printing—the file is uploaded in the printer software; the reservoir is filled with the ink; the ink can be preheated before printing; the printing parameters are defined, mainly for controlling the printing head (see next section); then, printing takes place.

5. Post printing—after printing the printed pattern must be sintered/cured (by temperature, photonic, UV light, plasma); a protective layer can be applied (e.g., by lamination, spray coating)

The IJP technology for PE applications and its technical-scientific developments have been reviewed over the last decades. Hue P. Le [44], 1999, reported the developments in the various IJP technologies, noting the significant growth rate of inkjet printers market. New ink formulations and new printhead designs were recognized as relevant for new applications. In 2010, the state-of-the-art of IJP of functional materials was reviewed by Raje and Murmu [45]. Improvements in process throughput remained the major challenge. In the same year, Derby reviewed the current understanding of the mechanisms of drop formation and the interactions between drops and the substrate, with a focus on the fabrication of structures for structural or functional materials applications [46]. Two years later, in 2012, Cummins and Desmulliez conducted a review in IJP of conductive materials [47]. IJP process, substrate properties, and types of conductive inks are the various factors that affect the quality of inkjet-printed products and their increasing relevance to the fields of electronics manufacturing, packaging, and assembly. In 2019, Nayak et al. reviewed the IJP of electronic devices, mainly addressing the fluid dynamics of inks and main properties (e.g., viscosity, surface tension, Weber number, Reynolds number, and Ohnesorge number) and their effects on defects appearance (coffee ring formation) [48]. The use of functional inks in sensors, thin-film transistors, and energy storage devices is presented. Ke Yan et al. revised the state-of-the-art related IJP strategies and functional inks for wearable electronic devices (e.g., sensors, displays, transistors, and energy storage devices) [49]. They highlighted the need of having available more intrinsically flexible and stretchable inks for avoiding, cracking, and delamination on highly flexible/stretchable substrates. Also, IJP technology development shall solve nozzle clogging issues for a more stable printing process. Kye-Si Kwon et al. reviewed piezo-driven IJP for PE. Other printing methods for high viscosity ink are also considered and compared (e.g., electrohydrodynamic jet, aerosol jet, and micro-plotter printing) [50]. There is a high demand for high-resolution printing of high viscosity inks for PE. In this case, the functionality of the device is more important than graphism perception, and the development of suitable inks for IJP remains one of the key issues. More recently, Muhammad Ali Shah reviewed the classifications and applications (textile, displays, and wearable devices) of IJP with more attention paid to piezoelectric IJP due to its higher relevance [36]. Various driving-voltage waveforms approaches are compared. Recently published studies on applications of IJP are summarized. Again, high high-viscosity IJP technologies are revised. The performance of IJP shall be improved by the development of new printheads with ink-recirculation and new techniques for printing high viscosity inks.

2.2 Functional inks for IJP

The use of inks has been around for almost as long as there is human life. It empowered evolution and was responsible for cultural and sociological developments whose footprints can be traced from the Paleolithic to this day [51]. The methods to dispense inks have also evolved with them and the first inkjet-type apparatus was patented in 1858 by William Thomson and Abbe Nollet [52]. The concept of printing functional conducive inks emerged some years later, in the 20th century, and was patented by Albert Hanson [53]. Nonetheless, research in IJP of functional inks exploded only nearly 100 years later, at the turn of the 21stcentury, thanks to the breakthrough development of organic conducting polymers by Heeger, MacDiarmid, and Shirakawa, which rendered them the 2000 Chemistry Nobel Prize [54, 55]. This led to several advancements in the field of PE, including the development of the first high-resolution printed all-polymer transistor circuits [56–58]. The fact that polymeric inks are more stable, easier to formulate, manipulate, and print was mostly responsible for this paradigm shift [43]. Nonetheless, with the technical developments experienced in this field, metallic-based inks started to be printed shortly after, thanks to the use of stable solvent systems and other additives that allowed to stabilize the metallic particles into homogeneously dispersed formulations with tunable surface tension and viscosity. Thereby, nowadays several base materials can be selected, depending on the final device desired functionality.

Functional inks for inkjet printing can be divided into, conductive, semi-conductive, and dielectric. Conductive inks are usually applied in the development of conductive tracks, vias, and electrodes. They generally rely on the dispersion of metallic nanoparticles, namely Ag [59], Cu [60], and Au [61], on organic or water-based solvents. To aid in the dispersion and grant long-term stability of these inks, surfactants, stabilizers, humectants, and other additive compounds are demanded [32]. To further tune the ink properties conductive nanofillers such as CNT can also be added. Although less conductive than metals, some polymers, metal-oxides, liquid-metal alloys, MXenes, perovskites, quantum-dots, and metal–organic-decomposition inks can also be used in the development of conductive inks. Currently, indium tin oxide (ITO) is still the most used material to produce transparent electrodes for thin-film devices (organic light-emitting diodes, OLED; field-effect transistors, OFET; photovoltaic devices, OPV) [62, 63]. However, the deposition o ITO is usually done by resorting to physical vapor deposition (PVD) which is much more expensive and energy-demanding than printing technologies. Moreover, its over-exploitation is damaging to the environment, and it is only recyclable through energy-consuming processes [64]. MXenes, quantum dots, and perovskites are examples of alternative base materials that can be used to develop inks for inkjet printing transparent electrodes for the above-mentioned applications. As a result, even though they have lower power conversion efficiency, their popularity is increasing [65, 66]. Despite being known for their higher electrical conductivity most inorganic inks are expensive, become brittle after curing, have limited flexibility, and might experience oxidation and loss of performance, if not properly encapsulated. As a result, organic conductive polymers such as poly(3, 4-ethylenedioxythiophene)-poly(styrenesulfonate) (PEDOT:PSS) [10, 67, 68], poly(3-hexylthiophene-2,5-diyl) (P3HT) [69, 70], and oxidized polyaniline (PANI) [71] are also being used as alternative materials for printing electrodes and conductive tracks. Other highly conductive inks can be developed using carbon allotropes, such as single-walled and multi-walled carbon nanotubes (SWCNT and MWCNT) [72–76], graphene [77, 78], and fullerenes [79].

Semiconductors have an electrical conductivity that can vary between the conductor and the dielectric. They can be n-type or p-type, depending on the doping atomic impurities added to the structure of the semiconductor. These impurities define the electrical properties, with highly doped semiconductors presenting conductivity values similar to metals. When the semiconducting material is less doped, its conductivity departs further from the conductive range. These semiconductors are crucial for the performance of the final device since their characteristics usually change with environmental physical, or chemical conditions [43, 72]. Less doped inorganic semiconductors include zinc-oxide (ZnO), zinc tin oxide (ZTO), and indium-zinc-oxide (IZO). Another interesting material for the development of semiconductor inkjet inks is the amorphous indium-gallium-zinc-oxide (a-IGZO), as it is processable at low temperatures, being vastly used in thin-film-transistors and solar cells [80]. On the other hand, organic semiconductors can be PEDOT: PSS, rubrene, pentacene, poly(diketopyrrolopyrrole-terthiophene) (PDPP3T), diphenylanthracene (DPA) [43]. PEDOT: PSS and CNT-based composites are vastly

Ink	Description	Examples
Conductive	Composed of highly conductive materials, mostly metals, metal-based composites, metal oxides, metallic alloys, highly doped conductive organic polymers in solvents, metal-oxide particles suspended in binders or organic-metallic blends	Metallic nanoparticle-based inks; Conductive polymers; Graphene inks; Perovskites; Mxenes; Quantum dots; Metal oxide-based inks; Eutectic liquid metals
Semi-conductive	Composed of semiconducting organic polyme*rs in solvents, inorganic nanoparticles (Group III-V, II-VI and IV semiconductors and carbon nanotubes) suspended in carrier fluids, or organic–inorganic blends; Also composed of less doped polymers that can be reinforced with carbon nanotubes or wires.	Metal-oxide inks; Semi-conductive polymers; Carbon-based inks
Dielectric	Organic polymers in solvents, organic polymer thermosets or ceramic-filled organic polymers	Ceramics; Metal–organic materials; Dielectric polymers; Piezoelectric polymers

Table 3.
Summary of ink types, their description, and some examples.

used as pressure sensors (piezorresistive materials) [81, 82], and temperature sensors (thermoresistive materials) [83, 84].

Finally, dielectrics exist in the less conductive boundary of the conductivity spectrum. They are used in electrical applications that demand high capacitance and insulation. Some dielectrics inks can be made from metal–organic materials such as aluminum oxide (Al_2O_3), zirconium oxide (ZrO_2), hafnium oxide (HfO_2), and yttrium oxide (YO2). Nevertheless, organic dielectric inks can also be formulated from polyvinylpyrrolidone (PVP), Polyvinyl alcohol (PVA), and Polymethyl methacrylate (PMMA), polydimethylsiloxane (PDMS), Polyvinylidene fluoride (PVDF) and its copolymer, polyvinylidene fluoride-trifluoroethylene (PVDF-TrFE). PVDF-based inks are of extreme importance since they are ferroelectric and enable piezoelectric applications. Another inkjet ink is the electrostrictive P(VDF-TrFE-CTFE) terpolymer, which can be used for energy harvesting applications [85–87]. **Table 3** summarizes the different types of printable inks.

2.3 Inkjet printing process variables

Both CIJ and DoD printing demand the use of inks with particle size under 1 μm (ideally <300 nm). In the particular case of the piezoelectric DoD method, their viscosity should be in the range of 1-20 cP, and their surface tension between 35 and 70 $mN.m^{-1}$ [35]. IJP requires no mask, has a low ink waste rate, and typical linewidth resolution of 30-50 μm [88]. DoD has established itself as the main IJP technology, with piezoelectric method being the most widely employed when it comes to IJP printing of functional inks, as it allows less ink consumption [49]. Hence, inks are specifically formulated to meet the requirements of the printing process [49].

To achieve the best possible printing quality, several factors need to be taken into account and studied from an optimization-driven perspective. For instance, the rheological properties of the ink (viscosity, surface tension, and density) interplay with each other and cannot be individually assessed. Similarly, the parameters of the printing process themselves cannot be individually studied. Thus, the printing resolution, printhead speed, printhead height, waveform profile, droplet size, and printhead temperature also influence one another and, as a result, before high-quality printing

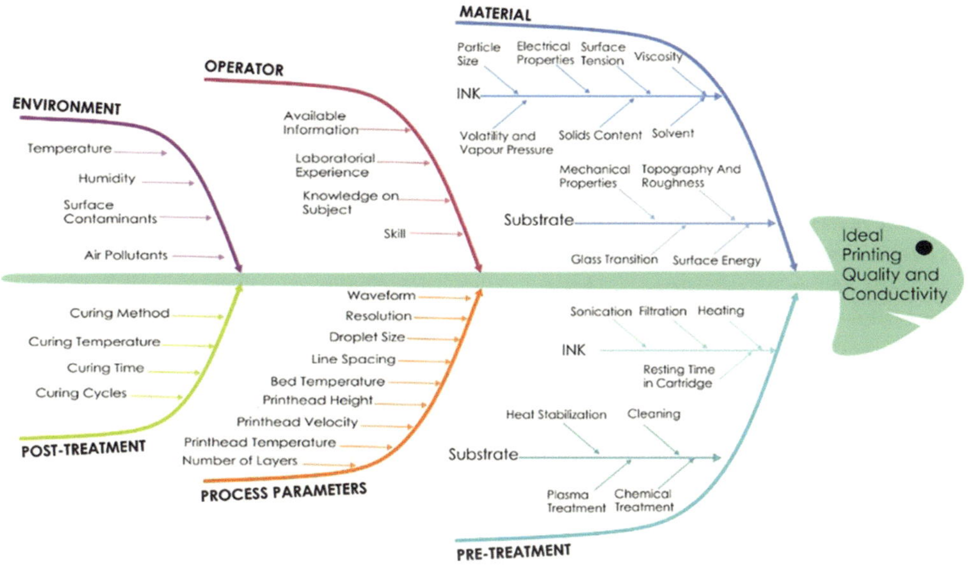

Figure 4.
Potential cause-effect factors influencing printing quality and conductivity.

can be attained, a series of optimization studies need to be conducted for each combination of ink/substrate. Other variables include the pre and post-treatment of the ink, substrate, and printed outputs. The operator experience and the environmental conditions will ultimately also influence the printing process. In **Figure 4**. the main print quality affecting parameters are summarized in a Fishbone diagram. Since several factors need to be simultaneously studied, a design of experiments approach is frequently performed by researchers aiming to rapidly optimize the process [59, 89–91].

Even after all parameters are optimized, some issues can still occur during the printing process namely nozzle clogging, printing deficiencies (coffee-ring effect, satellite drops, random electrical interference that causes droplet jetting oddness, missing droplets), loss of ink dispersibility, presence of dirt or dust particles in the ink system or substrate, among other issues [92]. As a result, to use IJP to develop electronic devices in industrial settings in-line quality control methods should be performed as a way of assuring functionality. To prevent the effect of environmental variables, which are often uncontrollable, the printing process should be performed in a controlled clean-room area.

2.3.1 Inks properties

Inks intended for inkjet printing should have linear Newtonian behavior and low viscosity, within a specific range. The drop formation and dynamics of the ink are ruled by three dimensionless numbers, that are related to the ink rheological and physical properties, namely the Reynolds number (R_e), the Weber number (W_e), and the Ohnesorge number (Oh) [49]:

$$\mathrm{R_e} = \frac{\upsilon \rho d}{\text{ç}};\mathrm{W_e} = \frac{\tilde{\text{o}}^2 \rho d}{\tilde{\text{a}}};Oh = \frac{\sqrt{\mathrm{W_e}}}{\mathrm{R_e}} = \frac{\text{ç}}{\sqrt{\gamma \rho d}};\mathrm{Z} = \frac{1}{Oh} \qquad (1)$$

where, υ is the velocity, ρ is the density, η is the dynamic viscosity, γ the surface tension, and d the characteristic length. To assure the ink is printable, its characteristics must obey some critical parameters and fall within certain limits. Most reports indicate that the optimum range to print a stable droplet is $1 < Z < 10$, as represented in **Figure 5** [49]. Z values above 10 relate to fluids with insufficient energy for drop formation, whereas Z values below 1 generally belong to fluids that are too viscous for printing, because the capillary force at the nozzle prevents its ejection. Also, to avoid the formation of satellite drops, the Weber number should be in the range of $2 < We < 25$. Nonetheless, these boundaries are not universal laws, with some authors reporting slightly broader ranges for ink printability [93, 94]. When the overall properties of the ink do not fall within these boundaries, the jetted fluid will not be able to form stable and consistent drops with adequate velocity to overcome the surface tension barrier at the tip of the nozzle, caused by the fluid/air interface. In this case, several break-up regimes can be identified from the Rayleigh break-up (insufficient velocity to form jets of ink) to the complete atomization (disintegration of the ink jets due to exaggerated velocity) [95].

2.3.2 Adhesion between inks and substrates

Another parameter that demands the previous study when planning inkjet printing is the adhesion between ink and substrate. This may affect the final printing result and additional procedures might be needed to assure compatibility. For ink to adhere well to a substrate, it must present appropriate wettability and adhesive bondability [96], meaning the surface tension of the ink must be lower than the wetting tension of the substrate (surface energy). To test this parameter, the substrate must be completely dry and free of any contaminants to start with. If after cleaning the substrate it still presents low wettability, surface pre-treatments may be needed. Such treatments include surface modification resorting to chemical modification,

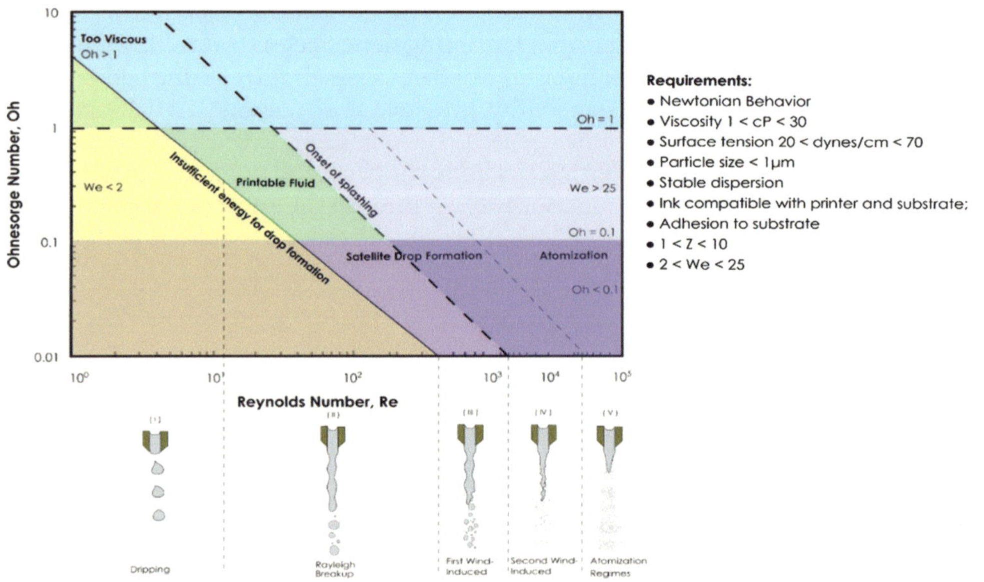

Figure 5.
Ohnesorge diagram, evidencing the fundamental characteristics of the printable fluid and the different drop break-up regimes responsible for printing errors, and requirements summary for developing a printable fluid.

coating with hydrophilic moieties (such as PVP and PVA solutions) [97, 98], plasma activation [98–100], mechanical abrasion, or superficial integration of nanoparticles or nanoclays [101], which increase surface roughness. Plasma activation is usually done by resorting to corona discharge plasma treatment and relies on the production of a high voltage electrical discharge that ionizes the molecules of the treated surface, positively changing its polarity (or oxidizing it) [99, 102, 103].

2.3.3 Pre- and post-treatments

To improve the printability of the ink some actions can be taken. Re-dispersion followed by filtration of the ink before use is effective in removing impurities and potential particle agglomerates that could clog the microscopic nozzles. The presence of trapped air bubbles in the ink is also damaging to the printing process. To prevent trapped bubbles, the ink should be degassed after filtering and left to rest for at least 30 minutes after filling the cartridge with it.

After printing, the ink must dry to become functional. When printing over paper and textiles, the ink is easily absorbed, and drying is not usually necessary, however, when using polymeric substrates it is preferable to promote the drying of the ink by thermal or UV curing [104]. In the case of thermal curing, heat is applied to facilitate the evaporation of the liquid ink carrier, which can be water or organic solvent. To render high printing quality and good electrical conductivity the curing temperature needs to be carefully chosen to prevent deformation, melting, or degradation of the substrate, as well as preserve the ink properties and avoid cracking [89]. To assure the homogeneous heating of the final printed patterns, an oven or an environmental chamber is usually used. Alternatively, cross-linkable inks are instantaneously cured using UV irradiation.

Regarding the drying of the ink itself, residual tensions might cause the ink not to dry homogeneously, leading to a phenomenon known as the coffee ring effect, which impairs the quality of the final print and is pictured in **Figure 6**. The coffee ring effect

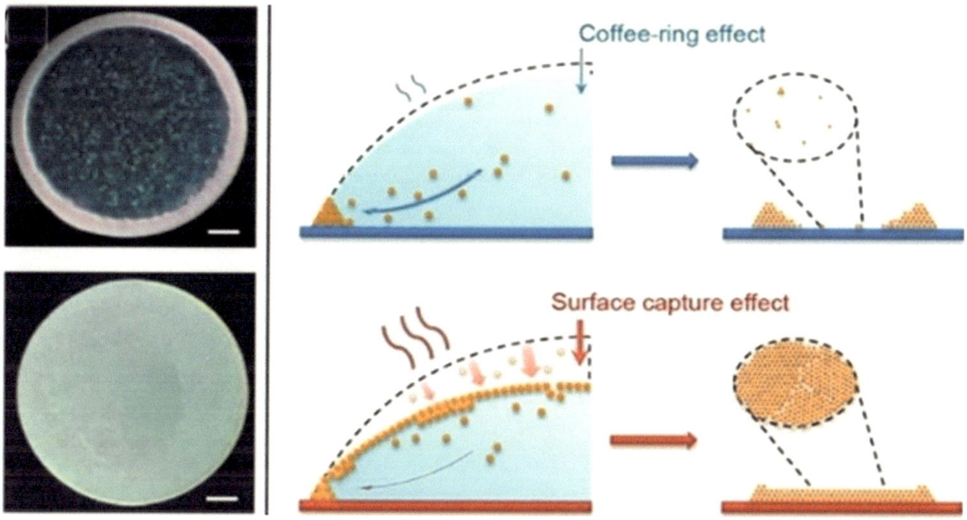

Figure 6.
Ring formation in colloidal droplets dried at room temperature versus uniform particle deposition of evaporating the same colloidal system in an environmental chamber at an elevated temperature. Reproduced with permission from refs. [105, 106]. Copyright © 2019 ACS.

is characterized by a ring-like morphology formed during ink evaporation, resulting from the solute segregation and accumulation along the drop periphery due to capillary flow. This issue occurs frequently when using nanoparticle inks and is caused by the convective macroscopic flow (Marangori flow) that occurs during the drying process and pushes the particles to the borders of the printed fluid [107], causing irregularities in thickness or coalescence between printed droplets.

Factors that influence the coffee ring effect are the evaporation rate and the particle concentration in the ink. A lower evaporation rate promotes ink homogeneity and can be reduced by lowering substrate temperature. Thus, the drying temperature and time also need to be optimized. Moreover, superficial cleaning, treatment, and heating must be performed homogeneously throughout the entire surface of interest. Since the ink starts its drying process as soon as it hits the substrate, the final curing step in the oven might not be sufficient to grant homogeneous drying. As a result, some printers encompass an integrated substrate heating feature, which can facilitate the bonding of the ink to the substrate and in between layers (when more than one layer of ink is printed). Using a solvent with a high boiling point, adding additives, reducing drop volume, and increasing the particle diameter can also help reduce the coffee ring effect.

2.3.4 Process and apparatus related parameters

Printhead model. Different inkjet printing systems demand different printhead models. The model of printhead also varies depending on the final application. Printheads for industrial IJP applications are more expensive, display a larger number of nozzles and allow for higher printing resolution. On the other hand, printheads intended for research and laboratory use are cheaper, have fewer nozzles, and print in lower resolution, as displayed in **Table 4** [50]. Regarding laboratory printing setups the most frequently used thorough the literature is the Dimatix Materials Printer, DMP series [50, 92, 108–112].

The general inkjet printer setup can be seen in **Figure 7a**. A 3-degree of freedom (DOF) system is the most frequently adopted and allows for efficient printing. Usually, the printhead only prints in one of the directions x-y directions (left-to-right).

In an inkjet printer the following parameters can be adjusted:

Resolution. The resolution should be chosen in accordance with the desired printed pattern, selected ink, and substrate material [109]. In multi-nozzle inkjet printers, the resolution is often controlled by the rotation of the printhead in predefined angles

		Features				
System Type	**Examples**	**Max. Jetting Frequency (kHz)**	**Print Width (mm)**	**Dropplet Volume (pL)**	**Resolution (dpi)**	**Number of nozzles**
Printheads for industrial applications	Xaar, Hitachi Ricoh, Konica Minolta, Kyocera	45–100	72–116	1.5–21	360–400	1024–5680
Laboratory and research systems	Microdrop,Microfab, Fujifilm Dimatix	30	64.96	12–33	1200	1024

Table 4.
Available inkjet printing systems [34, 50].

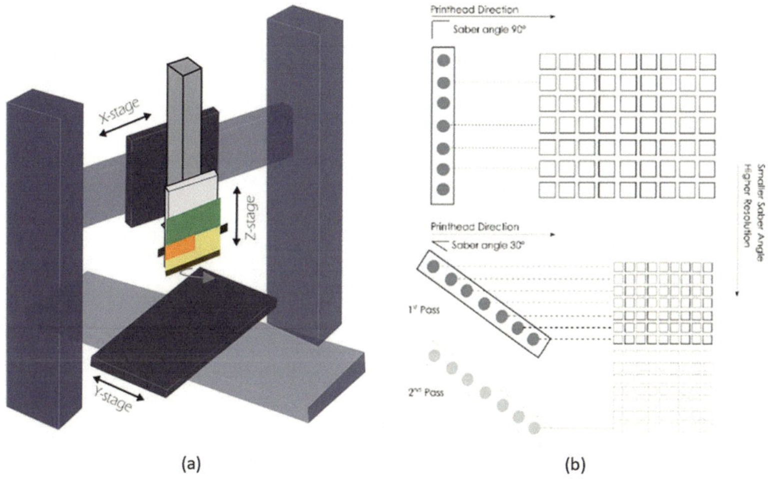

Figure 7.
(a) Illustration of the XYZ cartesian inkjet printing system with mounted printhead. (b) Depiction of the Printhead in its native position and rotated at a 30-degree angle [113].

(saber angles). To increase resolution, the printhead is rotated towards the movement direction axis. While the printhead is perpendicular to the printing direction, the resolution equals the native distance of the nozzles (lowest resolution). When the angular displacement is decreased, the distance between nozzles is diminished and the printed patterns achieve higher resolution (**Figure 7b**) [113]. In cases where the printhead is manually rotated, operator errors may occur, especially when dealing with small patterns with high resolution, which are more prone to displacement errors. For example, as seen in **Figure 7b**, many patterns demand more than one pass of the printhead be concluded and, if the direction is not perfectly aligned, the pattern will experience displacement between the two passes.

In practice, higher resolution is associated with a higher density of drops per area, which in turn, provides higher conductivity when printing conductive inks. Nonetheless, some issues might arise when the drops are too distanced (low resolution) or too overlapped (high resolution) and printing errors such as flooding, lack of superficial homogeneity, and loss of conductivity can occur, as illustrated in **Figure 8**. Moreover, if several passes are demanded additional attention must be given to assure correct angle alignment [109]. Some designs can also be prone to variability and errors, particularly if the image has sharp right-angle corners or lines whose width does not obey critical spacing rules [108].

Since inkjet printers usually print exclusively when moving in the left–right direction, different orientations of the same patterns may experience differences in the quality output, especially when higher resolutions are involved (> native resolution). This is particularly noticeable if the manual setting of the saber angle is not perfectly aligned. Considering this, to print the correct resolution in the precise position, the inkjet printer software is programmed to compensate for the angular displacement of the head. When this happens, even a slight imprecision during the printhead assembly can propagate printing errors across several printing passes (In **Table 5**, the major figures of merit that allow to identify the quality of printing are summarized, along with their description and observational examples).

Printhead Height. The height of the printhead, i.e., the distance between the nozzle and the substrate (the stand-off distance) can influence the splatter pattern

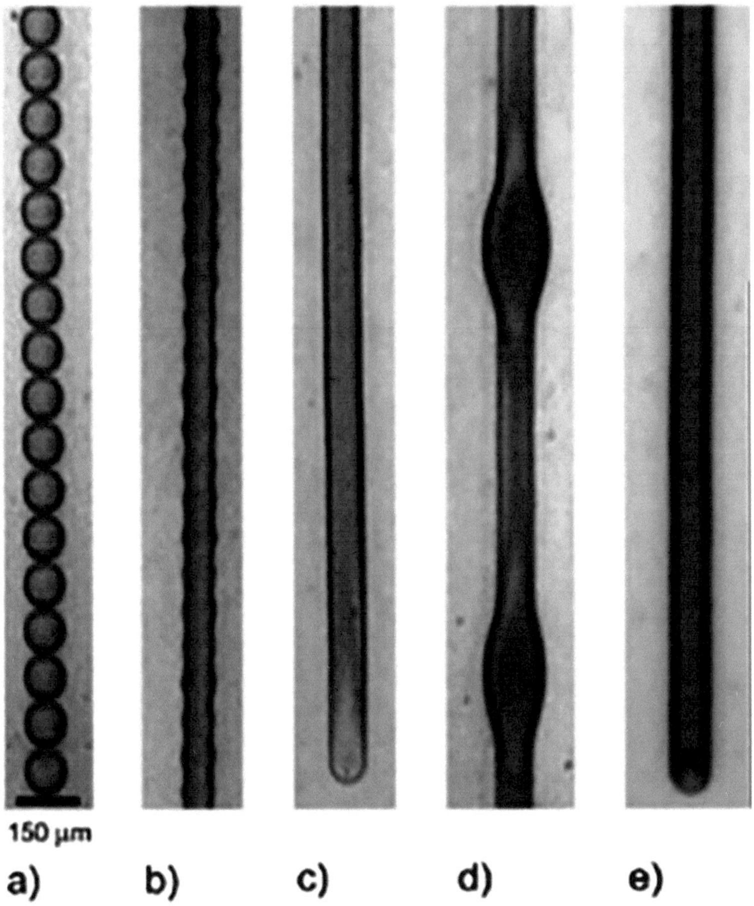

Figure 8.
Illustration evidencing the relationship between dot spacing and morphology of printed lines. (a) Large dot spacing caused drops to be isolated from each other. (b) Less drop spacing merges drops but their round edges are still visible. (c) Ideally merged drops forming a homogeneously printed line. (d and e) low drop spacing causing localized bleeding and coffee ring effect at the edges of the printed line, respectively. Reprinted with permission from [97]. Copyright (2022) ACS.

of the drops and their displacing accuracy. Because of this, higher printhead height is correlated with lower printing accuracy (higher prevalence of ink spraying and splatter). An optimum height should be found taking into consideration the type of ink used and the morphology the drops develop during the ejection phase. Some inks, especially metallic nanoparticle-based ones also demand heating of the printhead, as a way of helping the flow of the ink, as this promotes nanoparticle distancing and decreased viscosity, which avoids nozzle clogging [108, 109]. During printing, it is also essential to make sure the distance between the printhead and the substrate remains stable which in some printer models is assured through a vacuum-assisted bed that prevents substrate vibration and displacement [50].

Voltage Waveform. The waveform is the parameter that causes the piezoelectric crystal to deform, generating negative pressure at the printhead nozzles, and causing the ink drops to be jetted. To complete the drop ejection process, a bipolar voltage waveform is responsible for the fill/fire pulses (**Figure 9a**). The higher the voltage, the higher the speed of the generated droplet. This occurs because the pressure

Figures of merit	Description	Observations
Uniformity and homogeneity	Uniformity relates to ink thickness and width irregularities (e.g., line width, raggedness, blurriness). Homogeneity is linked to distribution of ink in the printed pattern area.	Coffee ring effect is an ink inhomogeneity feature. The size of a printed line depends on drop volume and droplet-substrate interactions. Typically, printed line width is higher that the design value. Ink thickness is measured by profilometry or confocal microscopy. Its morphology by optical or SEM. Image analyzer are used for quantitative characterization. Inconsistency due to misfiring, nozzle-plate flooding and satellite drop formation (related to the ink rheology and surface tension)
Consistency	Consistency of ink flow guaranteeing a constant process (drop volume, full dot drop)	
Resolution	Capacity to reproduce printed pattern details	Addressability (dpi, lpi of a inkjet printer or npi of a printhead); dot spacing; sharpness and contrast;
Drop placement error	Difference between the target and impact locations, related to printing accuracy and precision.	Error influenced by the precision of the nozzle manufacturing and the nozzle-substrate distance.
Repeatability	Agreement between successive measurements of drop placement	Can be tested through observation or electrical conductivity measurements (consistent for the same patterns over different prints)
Wettability	Preference of a fluid to maximize the surface contact with the substrate and spread over it	Weber number, Z number, contact angle, surface tension
Adhesion	Adhesion of the ink to the substrate and cohesion between layers	Surface tension, surface energy of the substrate, topography of substrate
Electrical resistivity	Can be measured as surface or volume resistivities	Two-point probes method or Van der Paw method (for better accuracy)
Electromechanical behavior	Deformability of the printed pattern	Gauge factor; geometrical layout of printed pattern
Ink fracture	Crackling of the deposited ink	Optical analysis
Flexibility/ stretchability	How much the ink is able to deform and return to its native state without losing its properties	Intrinsically flexible/stretchable inks; Flexible/ stretchable substrates; design of patterns to withstand repetitive stress loading

Table 5.
Summary of figures of merit used to classify the quality of the inkjet printing process.

created inside the chamber is consequently superior [116]. To recreate the desired image, the inkjet printer is connected to adequate computer software that supports the upload of the image and allows for the printing settings to be defined. In the case of the piezoelectric DoD method, a voltage is applied across a piezoelectric crystal under a pre-defined time and amplitude pattern, which generates the voltage

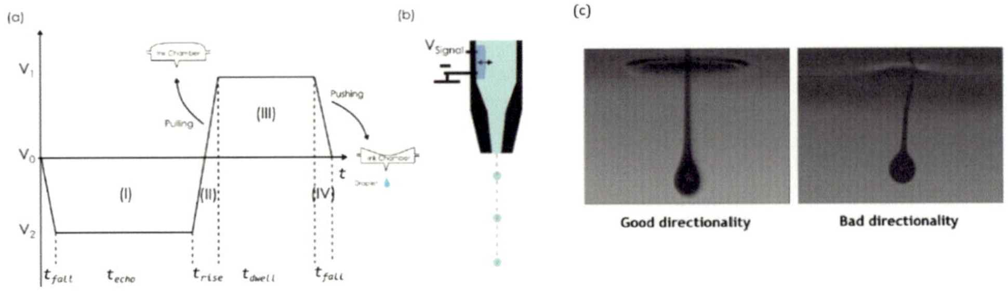

Figure 9.
(a) Simple depiction of a waveform. (I) Negative pulse that eliminates residual oscillations after each drop ejection; (II) and (III) are the pressurization and ejection phase, respectively; (b) nozzle pressure chamber as the piezoelectric crystal (darker blue) deforms due to step (III) [114, 115]; (c) jet straightness images obtained from a high-speed camera. Reprinted from [50].

waveform [114]. As depicted in **Figure 9a**, the waveform has four main phases – the damping and relaxation period that prepares the chamber to start the cycle (phase I, or t_{echo}), then, the pressurization (phase II) causes the ink chamber and nozzles to fill and defines the volume that will be ejected. During the next phase (phase III, or dwell time) there is a slight pause for stabilization of the ink inside the chamber. The cycle ends in phase IV, with the ejection of the ink droplets due to the pushing pressure caused by the piezoelectric crystal. Depending on the characteristics of the ink, the waveform can be optimized, and the best jetting performance is associated with fast, round, and stable droplets without tails [114]. To achieve this, the total waveform duration, number of phases, and the individual phase amplitude, durations, and slew rates (slopes), can be manually adjusted [114]. Recent inkjet printers are already equipped with waveform tuning ability, and in some cases encompass real-time drop-watcher high-speed cameras that capture images such as the ones in **Figure 9c** [108, 109]. This allows for simultaneous tracking and optimization of the profile of the jetted ink drops.

Jetting Frequency and Printhead Speed. The frequency in which the piezoelectric crystal is actuated is intimately related to the waveform and the printhead speed, which can be varied, affecting the rate at which droplets are jetted. If the printhead moves at a low speed the printhead nozzles will be actuated at a lower frequency. By generating waveforms with longer echo and dwell times, the jetting frequency will also be stalled, and vice-versa. As a result, this parameter can be simultaneously studied with the voltage and printhead speed to obtain higher quality printing. As seen in **Figure 10**, the shape of the jetted drops is dependent upon the jetting frequency and varies with the ink characteristic (viscosity, rheology, surface tension).

Number of printed layers. It is also a frequently studied parameter, particularly in what concerns its influence on electrical conductivity. To illustrate this, Rihen et al. studied the effects of the number of printed layers (1–5 layers), drop spacing (1016 DPI – 1693 DPI), and curing temperature (75–120°C) in the final conductivity of an Ag ink [89]. In this study, they found that the best conductivity was obtained for 3 printed layers, 1270 DPI, and a curing temperature of 120°C. It was also found that the number of printed layers strongly affected the final conductivity and that when too many layers were printed, excessive ink ejection ended up causing bleeding of the ink during printing and cracks after the heat-treatment was conducted. Hence, when possible, the number of printed layers should be limited to an "optimum minimum"

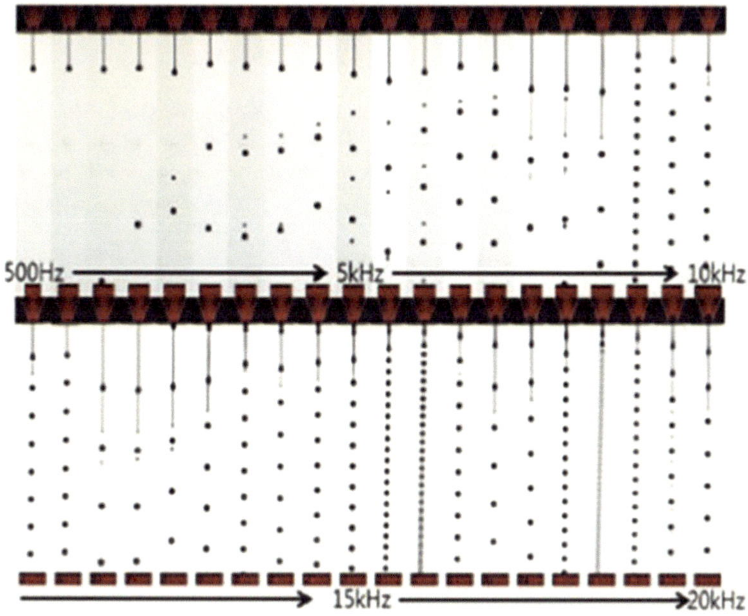

Figure 10.
Jetting behavior with increasing frequency. Reprinted from [117] with permission of AIP publishing.

as it can end up affecting conductivity and printing quality in a negative manner. Moreover, by finding a compromise between resolution and number of layers the printing process can become more economical in terms of ink used.

Design-related variables. The layout of the circuit designs to be printed needs to obey specific rules that depend upon the type of circuit, inkjet printer, printhead, and ink. The process of developing a certain circuit to be printed starts by defining the schematic circuit design in CAD, and after making sure it obeys the rules it is converted to bitmap for printing (**Figure 11a**) [118]. Design-related variables include the minimum resolution, horizontal/vertical line drop spacing, horizontal/vertical line width, horizontal/vertical line thickness, orientation of the design, and the angles of the line connections and design borders [118]. As depicted in **Figure 11c**, different line spacing between two consecutive lines can cause the lines to overlap. Hence, by managing the way the design is created, different results can be obtained depending on the final objective. Moreover, by anticipating the probability of flooding or coffee ring effect of the ink in specific areas, the design can be manipulated to prevent them through pattern compensation methods, as advanced by Vila et al. [119].

2.3.5 Inkjet printing quality indicators

Fabrication of PE devices using IJP faces a series of challenges for enhancing the technology merit indicators, which are application dependent. These relate mainly to printing quality indicators (printed pattern homogeneity, resolution, consistency), electrical conductivity, mechanical durability, and device flexibility/stretchability. In IJP, the printing quality involves complicated interactions between many factors including the printer, the printhead, the substrate, and the ink [49].

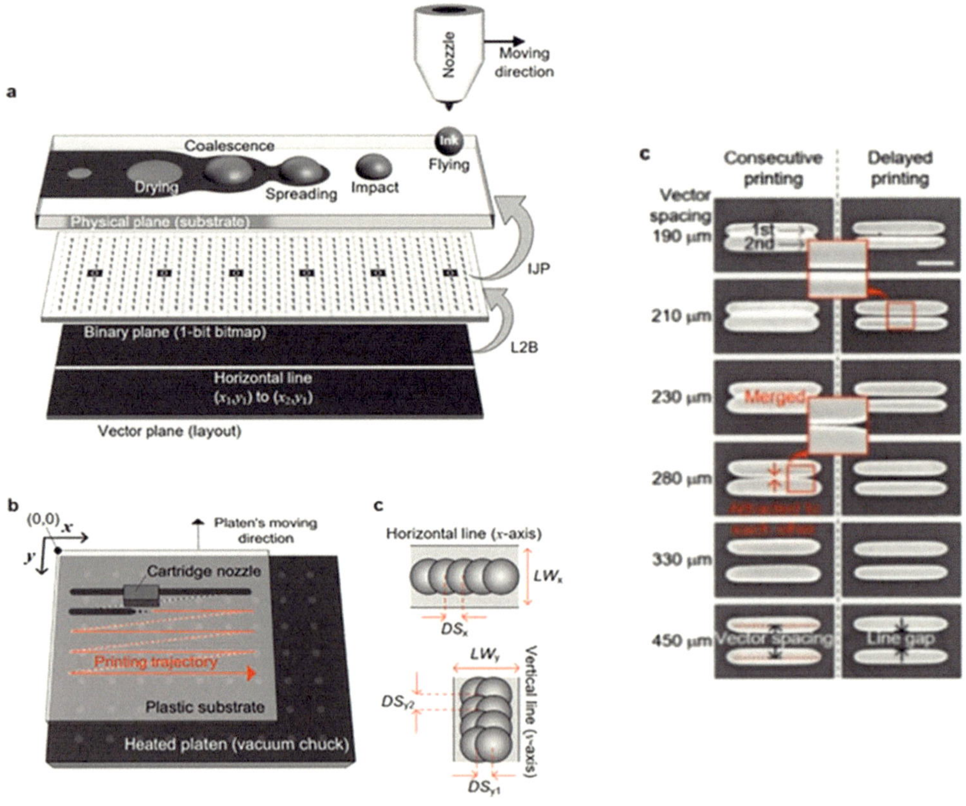

Figure 11.
(a) IJP metal track generation along the x-axis (mask-less): From a CAD layout drawing (a vector file) to a binary plane (a bmp file) and from the binary plane to a physical substrate surface. (b) a unidirectional IJP system. (c) Microscopic images of the two printed lines with varying vector spacings from 190 to 450 μm (scale bar is 500 μm).

Printed pattern resolution and uniformity are determined by droplet-substrate interactions, ink solvent evaporation rate, and capillary flow inside the ink droplet. The printed lines' dimensions, namely their width, depend upon the drop spacing and coalescing time. There is a relationship between drop spacing, line width, and electrical resistance. The electrical resistance is directly proportional to the drop spacing and inversely proportional to the line width.

3. Inkjet printing of smart products

IJP is already ubiquitously employed for printing decorative layers of products. In addition to this, during the past decade, it has also started to establish itself as a low-cost manufacturing process for large-area electronics applied to smart devices [43, 50]. As previously stated, potential markets include the development of electrodes and charge transport layers for thin-film devices, energy storage devices, electronic textiles, wearables, and smart tags and sensors for remote monitoring and logistics of marketable goods. In some of these cases, technological advancements have already allowed products to emerge and enter the market [120, 121].

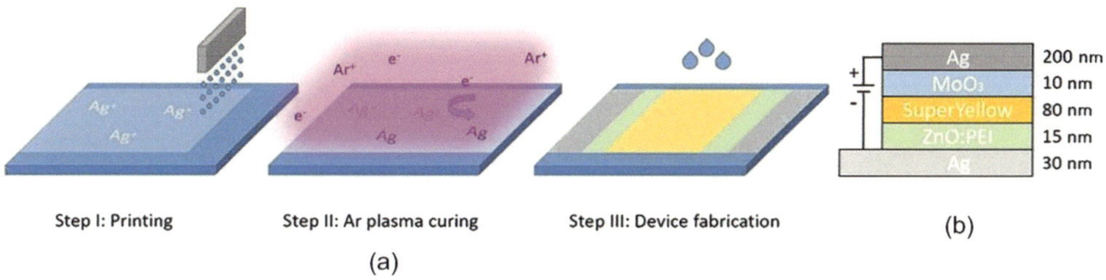

Figure 12.
Overview of the device fabrication. Reproduced from [123].

3.1 Flexible OLED and QLED

Many devices nowadays encompass screens made from transparent electrodes. These are usually obtained using metal oxides (such as ITO) and employed in light-emitting diodes (LED) devices. As stated above, however, ITO is brittle, and its exploitation and end-of-life cycle processing damaging to the environment. Hence, organic and hybrid alternatives, such as OLED and quantum-dot light-emitting diode (QLED) displays are starting to dominate the markets [122]. Thanks to the characteristics of the employed materials, inkjet printing is often a great manufacturing pathway to develop these devices [123]. As seen in **Figure 12**, ITO-free OLED can be obtained using inkjet-printed and low-temperature plasma-sintered Ag electrodes. A MOD ink was used to optimize the reduction effect of the plasma treatment, and the emissive layer of Super yellow was spin-coated.

Regarding organic printable materials capable of replacing the ITO electrodes, PEDOT:PSS is the favored organic semiconductor. Jürgensen et al. studied the tuning of a PEDOT:PSS solution with surfactants, as a way of inkjet printing green electrodes in OLED with reduced surface tension [124]. Moreover, Cinquino et al. concluded that by granting a surface tension value of28 – 40 mN/m and adding 40 vol.% of a low-boiling-point co-solvent proper substrate wetting was granted [125]. As for inorganic materials, perovskite nanocrystal (PeNC) solutions have also been investigated as inkjet printable color conversion layers (CCL) in PeNC/OLED hybrid displays [126]. These displays are used universally in entertainment devices, including augmented and virtual reality devices with enhanced performance. Another field of application of these displays respects healthcare devices and photomedicine, in particular, in the development of displays for photodynamic therapy (PDT), which vows to attack cancer cells using specific light-emitting wavelengths [127].

3.2 Organic photovoltaic (OPV) cells

Similarly, regarding OPV cells, the tendency is also to replace metal-oxide alloys with more environmentally friendly and easy to process materials. As an example, Alamri et al. developed fully-inkjet-printed hybrid perovskite photodetectors using Graphene/Perovskite/Graphene [128]. Schackmar and co-workers also came up with an approach to develop all-inkjet-printed absorbers and change transport layers [129]. As seen in **Figure 13a**), a p–i–n-perovskite solar cell architecture was created. The triple-cation perovskite absorber layer (TCP, brown) and

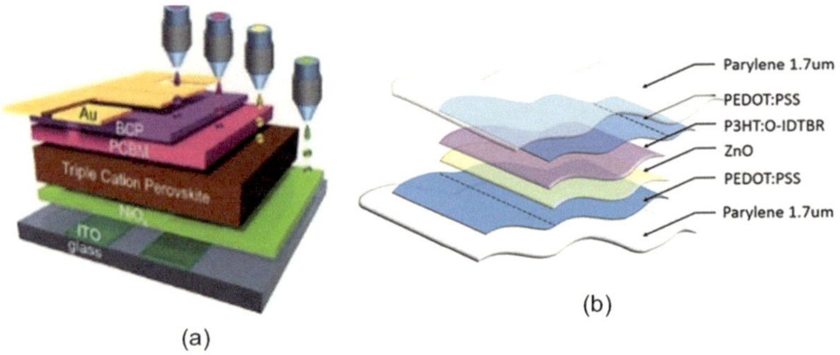

Figure 13.
(a) Schematic of the p–i–n-perovskite solar cell architecture with printed absorber and extraction layers. Reprinted from [129]; (b) schematic of the layer-by-layer composition of the solar cells made from PEDOT:PSS inkjet-printed electrodes. Reprinted from [69].

the double layer ETL made of PCBM and BCP (pink and purple, respectively) were deposited by inkjet printing. Bihar et al. also developed a fully-inkjet printed alternative to develop OPV in which PEDOT:PSS was used to develop the electrodes (**Figure 13b**) [69].

3.3 Energy storage applications

Inkjet printing technology is being vastly employed in the development of supercapacitors (SC), triboelectric nanogenerators, and batteries. Even though many challenges still have to be overcome for these devices to reach competitive performance, promising alternatives already exist [130]. For example, graphene-based solutions are vastly studied throughout the literature for IJP of supercapacitors [131, 132]. Li et al. inkjet printed disposable micro-supercapacitors (MSC) on paper using conductive inks based on the ternary composite of PEDOT:PSS, graphene quantum dots, and graphene [133]. In **Figure 14**, the resulting MSC are pictured in different array dispositions.

Giannakou and colleagues developed 3D conformable supercapacitors intended for epidermal energy storage. To achieve this, they inkjet-printed nickel (II) oxide active electrodes over PVA substrate. As a proof-of-concept, the SC was used on the skin of test subjects, and energy from their movements was successfully harvested to light a LED [134]. Energy harvesters can also be printed over textiles as a way of obtaining self-powered garments with sensing and monitoring abilities [135].

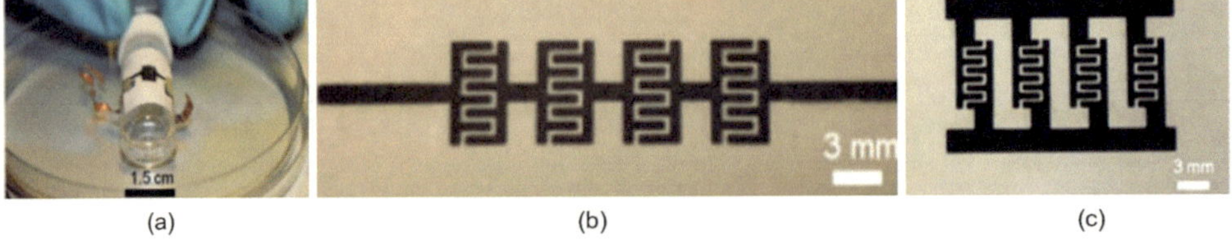

Figure 14.
(a) Microsupercapacitor (MSC) printed on photo paper for flexibility performance test, (b) fully-printed MSC array with 4 MSC connected in series on carton paper, (c) fully-printed MSC array with 4 MSC connected in parallel on carton paper. The bus lines were printed with 17 passes. Reprinted from [133].

VARTA company has recently started to apply inkjet printing to the development of batteries to power sensors, and smart tags for intelligent packaging applications. Different electrochemical systems and electrodes can be printed in a stacked or coplanar manner depending on the envisioned design of the battery [136].

3.4 Sensors, e-textiles, and biomonitoring devices

Sensors are vital to transduce physical changes into readable data. Several inkjetable materials can be used as the functional part of sensors, whose electrical conductivity varies according to those changes and is later processed into digital outputs for monitoring. The most frequently developed physical sensors measure mechanical (pressure, force, strain), temperature, and humidity changes. Metal nanowires [137], metallic nanoparticles [138], polymer micro/nanostructures [139], CNT, and graphene have been applied to the design of piezoresistive flexible tactile sensors [140–142]. To work efficiently, the latter ones must be homogeneously dispersed in an elastomeric matrix, in concentrations above the electrical conductivity percolation threshold [143]. To produce piezoelectric pressure sensors the most used materials are piezoelectric ceramics, ceramic/polymer composites, and single crystals [144, 145]. As for capacitive sensing applications, SWNT/PDMS electrodes are effective options [145, 146]. Inkjet printing has also been extensively used to produce temperature, and humidity sensors that can be applied in standalone settings or, thanks to the development of the IoT can work as scattered sensor networks for remote and connected monitoring applications. Thanks to their inherent conformability, low-cost, biocompatibility, tunability, accuracy, and adequate sensing range, the pressure, temperature, and humidity printable sensors, have started to be applied in e-textiles and biomonitoring applications. As an example, Farooqui et al. successfully developed a smart bandage to remotely monitor chronic wounds through inkjet printing of a resistive sensor sensitive to pH [147].

Wearables and electronic textiles can be used for applications ranging from human-machine interaction (HMI), fashion, haptics, and biomonitoring. Regarding biomonitoring, different sensors can be inkjet-printed over textiles or conformable polymeric substrates (PDMS, PET, PEN, PEEK) and retrieve accurate biological data, thanks to the close proximity to the body. Pressure, strain [67], temperature [148], and humidity sensors [17], are the most frequently printed, nonetheless, photoplethysmography (PPG), electrocardiography (ECG), and electroencephalogram (EEG)

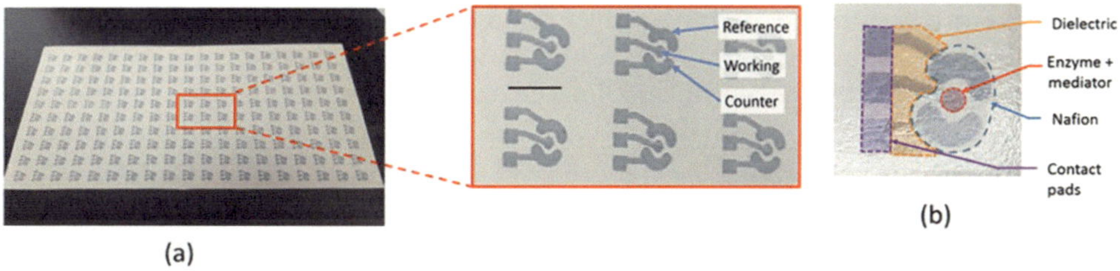

Figure 15.
(a) Inkjet-printed glucose biosensors; (b) fully printed biosensor and identification of the different printed layers, namely the electrode (PEDOT:PSS), the dielectric, the biological coating containing the enzyme and the mediator, and the encapsulation layer. Reprinted from [156].

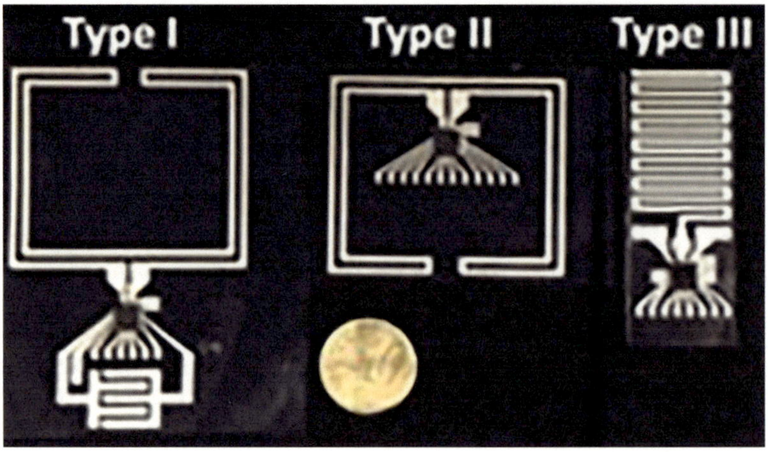

Figure 16.
Three types of printed, passive tags on a flexible substrate for operation in the UHF RFID band (902–928 MHz). Reprinted from [161].

sensors can be inkjet-printed as well [49, 135, 149–152]. Electroluminescent devices can also be printed over textiles to enhance their functionality [153].

Flexible printed and biocompatible sensors placed in direct contact with the human body are also valuable for sensing specific biomarkers. This can be achieved by IJP of enzyme-functionalized inks [154]. Mass et al. enzyme-functionalized silica nanoparticles and mixed them with SWCNT to create a bio-ink with catalytic activity [154]. Biocompatible graphene-based biosensors can be printed as well to monitor the effect of antiviral drugs through impedance analysis [155]. Bihar and colleagues also developed a disposable glucose sensor by inkjet printing PEDOT:PSS as electrodes, and a glucose oxidase solution as the sensing material [156]. A dielectric ink was printed to isolate the electrode interconnects as depicted in **Figure 15**.

3.5 Smart tags and logistics

Inkjet printing can be used in the development of antennas, radio frequency identifier (RFID) chips, and near-field communication (NFC) chips, which can work as smart labels and sensor tags. Temperature, humidity, and strain sensors are usually paired with these labels to develop intelligent packaging and/or tracking applications [157, 158]. For this purpose, paper is one of the most used substrates [159]. Another important application for smart tags is food quality monitoring. By combining humidity, ammonia, temperature, and volatile organic compounds (VOC) sensors the state of perishable goods can be evaluated and the food supply chain optimized accordingly [160]. For this purpose, Quintero et al. developed a multi-sensing platform where an RFID chip was integrated with inkjet-printed sensors (ammonia, humidity, and temperature) over a PEN substrate [160]. Baubauer and co-workers also studied the printing of different types of passive tags over flexible substrates, when integrated with a rigid RFID chip, as illustrated in **Figure 16** [161]. In this case, the purpose of the tags was to serve as user-interactive touch sensors. One interesting asset of these types of labels is the fact that they can be reset and reprogrammed.

Since packaging is meant to be disposable, by recovering and reprogramming the tags they can be reused in other applications before being ultimately recycled [162].

4. Advanced inkjet printing techniques

IJP is a mature technology and recently has been used to print functional inks for PE devices. This novel use demanded developments in terms of new printing equipment and inks. As a result, advanced IJP technologies have emerged, responding to the requirements of novel applications. These advanced technologies are focused on increasing printing resolution and speed, printing of high viscosity inks (with higher electrical conductivity), printing over non-planar substrates, and enlarging the range of materials that can be printed.

A novel double-shot IJP technique has been developed, which allows for the deposition of two types of inks at the same position [163]. In this way, conductive and dielectric inks can be printed at the same position, allowing the construction of devices. Reactive inkjet printing, which also uses two nozzles, combines the processes of material deposition and chemical reaction to print over a substrate material, enlarging the type of ink materials.

EHD-IJP allowed high-resolution printing, paving its use in micro/nano manufacturing of electronic devices [45]. EHD-IJP is a direct patterning technique that can also be used as a thin film deposition technique (e.g., electrospraying, electrospinning). Furthermore, multi-nozzle implementation has been proposed, but nozzle density is still low [50]. EHD-IJP also allows printing of high viscosity inks and consequently has huge potential for fabricating 3D patterns [163].

Needle-based printing is a recent technology to dispense relatively high viscosity ink through a fine nozzle [50]. Droplets are ejected from the nozzle exit by the motion of the needle, which can be operated by air pressure or piezoelectric actuator. This technique can handle high viscosity inks.

Micro-plotter is a technology that also allows dispensing of relatively high viscosity ink through a fine nozzle [50]. The dispensing mechanism is based on the ultrasonic pumping action at the core of the micro-plotter head, a micropipette. The mechanism is capable of depositing ink droplets with dot size of less than 2 μm, which is smaller than an inkjet system, even when using relatively high viscosity inks (up to 450 cP).

Other droplet-based techniques have been developed, mainly for printing on non-planar substrates, such as aerosol jet printing, surpassing the limitations of both inkjet and EHD-IJP techniques. Aerosol jet printing uses high-speed ejection of aerosols instead of liquid droplets [164, 165]. The aerosol is produced by atomization of ink, which produces very small droplets with diameters in the range of 1 to 5 μm. Due to the aerodynamic effect, it also allows higher resolution for fabricating micro- and nanoscale devices. Furthermore, aerosol jetting allows different inks to be conformably printed onto the substrate.

High-resolution 3D patterning combines IJP with 3D printing technologies [166]. High-resolution insulating and conductive layers can be printed using multiple printheads in the same printing system. However, this requires a high degree of deposition precision that can be achieved by the use of phase-change inks (activated by chemical or thermal triggers) with no solvents. 3D structures can also be produced by IJP, by deposition of layer-by-layer of two reactive components, followed by polymerization. This is a technology of current intense research and fast growth.

5. Concluding remarks and future trends

In this chapter, the main topics concerning the IJP manufacturing of printed electronics have been discussed. Being an additive manufacturing technology, some of its advantages concern lower associated costs, material economization, and design freedom. Since its establishment as an alternative for electronics manufacturing, different variations of this technology have emerged, nonetheless, piezoelectric IJP is the most widespread method. Many functional inks have already been optimized for IJP and are commercially available. Despite this, several factors linked to the printing process can still negatively influence the printing output in terms of both printing quality and electrical conductivity and demand optimization.

As for IJP application in the current PE market, there is an undeniable growing tendency. However, it is still early to anticipate the potential of IJP technology in the long run because of all the other options currently being developed. Because of the drawbacks associated with piezoelectric IJP, namely the tendency for nozzles to clog, the limited ink viscosity range (and consequently limited electrical conductivity), and the elevated number of factors that influence the printing process, it is possible that IJP gets to be used in specific niche applications, whereas needle-based and EHD-IJP might ultimately transcend piezoelectric IJP in terms of applicability.

Acknowledgements

This work has been supported by NORTE-06-3559-FSE-000018, integrated invitation NORTE-59-2018-2041, aiming Hiring of Highly Qualified Human Resources, co-financed by the Regional Operational Programme North 2020, thematic area of Competitiveness and Employment, through the European Social Fund, and by the scope of projects with references UIDB/05256/2020 and UIDP/05256/2020, financed by FCT—Fundação para a Ciência e Tecnologia, Portugal.

Author details

Cláudia Buga[1,2] and Júlio C. Viana[1,2]*

1 IPC/LASI—Institute for Polymers and Composites/Associated Laboratory for Smart Systems, University of Minho, Guimarães, Portugal

2 DTx Colab—Digital Transformation, University of Minho, Guimarães, Portugal

*Address all correspondence to: jcv@dep.uminho.pt

References

[1] Cruz SMF, Rocha LA, Viana JC. Printing technologies on flexible substrates for printed electronics. In: Rackauskas S, editor. Flexible Electronics. Rijeka: IntechOpen; 2018. DOI: 10.5772/intechopen.76161

[2] Khan S, Lorenzelli L, Dahiya RS. Technologies for Printing Sensors and Electronics over Large Flexible Substrates: A review. IEEE Sensors Journal. 2015;**15**(6):3164-3185

[3] Matsui H, Takeda Y, Tokito S. Flexible and printed organic transistors: From materials to integrated circuits. Organic Electronics. 2019;**75**:105432

[4] Correia V, Mitra KY, Castro H, Rocha JG, Sowade E, Baumann RR, et al. Design and fabrication of multilayer inkjet-printed passive components for printed electronics circuit development. Journal of Manufacturing Processes. 2018;**31**:364-371. DOI: 10.1016/j.jmapro.2017.11.016

[5] Castro HF, Correia V, Sowade E, Mitra KY, Rocha JG, Baumann RR, et al. All-inkjet-printed low-pass filters with adjustable cutoff frequency consisting of resistors, inductors and transistors for sensor applications. Organic Electronics. 2016;**38**:205-212

[6] Correia V, Oliveira J, Perinka N, Costa P, Sowade E, Mitra KY, et al. All-printed Piezoresistive sensor matrix with organic thin-film transistors as a switch for crosstalk reduction. ACS Applied Electronic Materials. 2020;**2**(5):1470-1477. DOI: 10.1021/acsaelm.0c00214

[7] Franco M, Alves R, Perinka N, Tubio C, Costa P, Lanceros-Mendéz S. Water-based graphene inks for all-printed temperature and deformation sensors. ACS Applied Electronic Materials. 2020;**2**(9):2857-2867. DOI: 10.1021/acsaelm.0c00508

[8] Yap YL, Wang C, Sing SL, Dikshit V, Yeong WY, Wei J. Material jetting additive manufacturing: An experimental study using designed metrological benchmarks. Precision Engineering. 2017;**50**:275-285. DOI: 10.1016/j.precisioneng.2017.05.015

[9] Krykpayev B, Farooqui MF, Bilal RM, Shamim A. A WiFi tracking device printed directly on textile for wearable electronics applications. In: IEEE MTT-S International Microwave Symposium Digest. 2016. p. 4

[10] Ma S, Ribeiro F, Powell K, Lutian J, Møller C, Large T, et al. Fabrication of novel transparent touch sensing device via drop-on-demand inkjet printing technique. ACS Applied Materials & Interfaces. 2015;**7**(39):21628-21633

[11] Ando B, Baglio S, Castorina S, Crispino R, Marletta V. An inkjet printed pressure sensor for applications in active ageing monitoring. In: SAS 2019-2019 IEEE Sensors Applications Symposium, Conference Proceedings; 2019

[12] Dankoco MD, Tesfay GY, Benevent E, Bendahan M. Temperature sensor realized by inkjet printing process on flexible substrate. Materials Science & Engineering, B: Solid-State Materials for Advanced Technology. 2016;**205**:1-5

[13] Vuorinen T, Niittynen J, Kankkunen T, Kraft TM, Mäntysalo M. Inkjet-printed graphene/PEDOT:PSS temperature sensors on a skin-conformable polyurethane substrate. Scientific Reports. 2016;**6**(1):35289. DOI: 10.1038/srep35289

[14] Katerinopoulou D, Zalar P, Sweelssen J, Kiriakidis G, Rentrop C, Groen P, et al. Large-area all-printed temperature sensing surfaces using novel composite thermistor materials. Advanced Electronic Materials. 2019;**5**(2):1-7

[15] Popov VI, Kotin IA, Nebogatikova NA, Smagulova SA, Antonova IV. Graphene-PEDOT: PSS humidity sensors for high sensitive, low-cost, highly-reliable, flexible, and printed electronics. Materials. 2019;**12**(21):1-9

[16] Jung M, Lee J, Vishwanath SK, Kwon OS, Ahn CW, Shin K, et al. Flexible multimodal sensor inspired by human skin based on hair-type flow, temperature, and pressure. Flexible and Printed Electronics. 2020;**5**(2):16

[17] Barmpakos D, Segkos A, Tsamis C, Kaltsas G. A disposable flexible humidity sensor directly printed on paper formedical applications. Journal of Physics Conference Series. 2017;**931**(1): 012003

[18] Cruz S, Viana JC, Dias D, Rocha LA. Pressure sensing platform for health monitoring. In: 2014 IEEE International Symposium on Medical Measurements and Applications (MeMeA); 2014. pp. 1-5

[19] Salim A, Lim S. Review of recent inkjet-printed capacitive tactile sensors. Sensors. 2017;**17**(11):2593

[20] Chiolerio A, Rivolo P, Porro S, Stassi S, Ricciardi S, Mandracci P, et al. Inkjet-printed PEDOT:PSS electrodes on plasma-modified PDMS nanocomposites: Quantifying plasma treatment hardness. RSC Advances. 2014;**4**(93):51477-51485. DOI: 10.1039/C4RA06878E

[21] Mohammed MG, Kramer R. All-printed flexible and stretchable electronics. Advanced Materials. 2017;**29**(19):1604965. DOI: 10.1002/adma.201604965

[22] Seipel S, Yu J, Periyasamy AP, Viková M, Vik M, Nierstrasz VA. Characterization and optimization of an inkjet-printed smart textile UV-sensor cured with UV-LED light. IOP Conf Ser. Materials Science and Engineering. 2017;**254**(7):072023

[23] Gardner SD, Haider MR, Islam MT, Alexander JID, Massoud Y. Aluminum-doped zinc oxide (ZnO) inkjet-printed piezoelectric Array for pressure gradient mapping. Midwest Symposium on Circuits and Systems. 2019;**2019**:1101-1104

[24] Fang M, Li T, Zhang S, Rao KV, Belova L. Design and tailoring of inks for inkjet patterning of metal oxides. Royal Society Open Science. 2020;**7**(4):200242. Available from: https://royalsocietypublishing.org/doi/abs/10.1098/rsos.200242

[25] Printed Electronics Market 2021: Industry Size, Regions, Emerging Trends, Growth Insights, Development Scenario, Opportunities, and Forecast By 2027 [Internet]. 2021. Available from: https://www.marketstatsville.com/printed-electronics-market

[26] Functional Printing-Global Market Trajectory and Analytics. 2021

[27] Inkjet Printing Market-Growth, Trends, Covid-19 Impact, and Forecasts (2022-2027) [Internet]. 2021. Available from: https://www.mordorintelligence.com/industry-reports/inkjet-printing-market

[28] Functional Inks-Global Market Trajectory and Analytics [Internet]. 2021. Available from: https://www.researchandmarkets.com/reports/5302023/

functional-inks-global-market-trajectory-and

[29] Global Printed Electronics Market-Industry Trends and Forecast to 2028 [Internet]. 2021. Available from: https://www.databridgemarketresearch.com/reports/global-printed-electronics-market

[30] Global Functional Printing Market-Industry Trends and Forecast to 2027 [Internet]. 2020. Available from: https://www.databridgemarketresearch.com/reports/global-functional-printing-market

[31] IDTechEx. Materials for Printed/Flexible Electronics 2021-2031: Technologies, Applications, Market Forecasts. 2021. IDTechEx Research finds, the most interesting emerging technologies

[32] Cruz SMF, Rocha LA, Viana JC. Printing technologies on flexible substrates for printed electronics. Rackauskas S, editor. Flexible Electronics. London: IntechOpen; 2018. p. 25

[33] Rivadeneyra A, Loghin FC, Falco A. Technological integration in printed electronics. In: Rackauskas S, editor. Flexible Electronics. Rijeka: IntechOpen; 2018. DOI: 10.5772/intechopen.76520

[34] Beedasy V, Smith PJ. Printed electronics as prepared by inkjet printing. Materials. 2020;**13**(3):704. Available from: https://pubmed.ncbi.nlm.nih.gov/32033206

[35] Sattler KD. 21st Century nanoscience—A handbook: Design strategies for synthesis and fabrication Sattler KD, editor. Vol. 1st ed. United Kingdom: Taylor & Francis; 2019. p. 496

[36] Shah M, Lee D-G, Lee B-Y, Hur S. Classifications and applications of inkjet printing technology: A review. IEEE Access. 2021;(9):140079-140102

[37] Kwon H, Hong J, Nam SY, Choi HH, Li X, Jeong YJ, et al. Overview of recent progress in electrohydrodynamic jet printing in practical printed electronics: Focus on the variety of printable materials for each component. Advanced Materials. 2021;**2**(17):5593-5615. DOI: 10.1039/D1MA00463H

[38] Yu M, Ahn KH, Lee SJ. Design optimization of ink in electrohydrodynamic jet printing: Effect of viscoelasticity on the formation of Taylor cone jet. Materials and Design. 2016;**89**:109-115. Available from: https://www.sciencedirect.com/science/article/pii/S0264127515305505

[39] Cai S, Sun Y, Wang Z, Yang W, Li X, Yu H. Mechanisms, influencing factors, and applications of electrohydrodynamic jet printing. Nanotechnology Reviews. 2021;**10**(1):1046-1078. DOI: 10.1515/ntrev-2021-0073

[40] Skarżyński K, Krzemiński J, Jakubowska M, Słoma M. Highly conductive electronics circuits from aerosol jet printed silver inks. Scientific Reports. 2021;**11**(1):18141. DOI: 10.1038/s41598-021-97312-5

[41] Modak CD, Kumar A, Tripathy A, Sen P. Drop impact printing. Nature Communications. 2020;**11**(1):4327. DOI: 10.1038/s41467-020-18103-6

[42] Foresti D, Kroll K, Amissah R, Sillani F, Homan K, Poulikakos D, et al. Acoustophoretic printing. Science Advances. 2018;**4**:eaat1659

[43] Buga CS, Viana JC. A review on materials and technologies for organic Large-area electronics. Advanced Materials Technologies. 2021;**6**(6):2001016. DOI: 10.1002/admt.202001016

[44] Le HP. Progress and trends in ink-jet printing technology. Journal of Imaging Science and Technology. 1998;**42**:49-62

[45] Raje P, Murmu NC. A review on electrohydrodynamic-inkjet printing technology. International Journal of Emerging Technology and Advanced Engineering. 2014;**4**(5):174-183

[46] Derby B. Inkjet printing of functional and structural materials: Fluid property requirements, feature stability, and resolution. Annual Review of Materials Research. 2010;**40**(1):395-414. DOI: 10.1146/annurev-matsci-070909-104502

[47] Cummins G, Desmulliez MPY. Inkjet printing of conductive materials: A review. Circuit World. 2012;**38**(4):193-213. DOI: 10.1108/03056121211280413

[48] Nayak L, Mohanty S, Nayak SK, Ramadoss A. A review on inkjet printing of nanoparticle inks for flexible electronics. Journal of Materials Chemistry C. 2019;7(29):8771-8795. DOI: 10.1039/C9TC01630A

[49] Yan K, Li J, Pan L, Shi Y. Inkjet printing for flexible and wearable electronics. APL Materials. 2020;**8**(12):120705. DOI: 10.1063/5.0031669

[50] Kwon K-S, Rahman MK, Phung TH, Hoath S, Jeong S, Kim JS. Review of digital printing technologies for electronic materials. Flexible and Printed Electronics. 2020;**5**:043003

[51] Halverson J. The first pictures: Perceptual foundations of Paleolithic art. Perception. 1992;**21**(3):389-404. DOI: 10.1068/p210389

[52] Wijshoff H. Structure and fluid dynamics in piezo inkjet printheads. Integrated Assessment. The Netherlands: Univ. Twente, Enschede; 2008

[53] Hanson A. British Patent 4.681; 1903

[54] The Nobel Prize in Chemistry 2000. [cited 2020 Nov 28]. Available from: https://www.nobelprize.org/prizes/chemistry/2000/summary/

[55] Brédas JL, Marder SR, Salaneck WR, Heeger AJ, MacDiarmid AG, Shirakawa H. Macromolecules. 2002;**35**(4):1137-1139. DOI: 10.1021/ma0118973

[56] Fan H, Lu Y, Stump A, Reed ST, Baer T, Schunk R, et al. Rapid prototyping of patterned functional nanostructures. Nature. 2000;**405**(6782):56-60. DOI: 10.1038/35011026

[57] Sirringhaus H, Kawase T, Friend RH, Shimoda T, Inbasekaran M, Wu W, et al. High-resolution inkjet printing of all-polymer transistor circuits. Science. 2000;**290**(5499):2123-2126. Available from: https://www.science.org/doi/abs/10.1126/science.290.5499.2123

[58] Evans JRG. Direct ink jet printing of ceramics: Experiment in teleology. British Ceramic Transactions. 2001;**100**(3):124-128. DOI: 10.1179/096797801681332

[59] Mypati S, Dhanushkodi SR, McLaren M, Docoslis A, Peppley BA, Barz DPJ. Optimized inkjet-printed silver nanoparticle films: Theoretical and experimental investigations. RSC Advances. 2018;**8**(35):19679-19689. DOI: 10.1039/C8RA03627F

[60] Li W, Li L, Gao Y, Hu D, Li C-F, Zhang H, et al. Highly conductive copper films based on submicron copper particles/copper complex inks for printed electronics: Microstructure, resistivity, oxidation resistance, and long-term stability. Journal of Alloys and Compounds. 2018;**732**:240-247. Available from: https://www.sciencedirect.com/science/article/pii/S0925838817336290

[61] Cui W, Lu W, Zhang Y, Lin G, Wei T, Jiang L. Gold nanoparticle ink suitable for electric-conductive pattern fabrication using in ink-jet printing technology. Colloids and Surfaces A: Physicochemical and Engineering Aspects. 2010;**358**(1):35-41. Available from: https://www.sciencedirect.com/science/article/pii/S0927775710000403

[62] Kim MG, Kanatzidis MG, Facchetti A, Marks TJ. Low-temperature fabrication of high-performance metal oxide thin-film electronics via combustion processing. Nature Materials. 2011;**10**(5):382-388. DOI: 10.1038/nmat3011

[63] Que M, Guo W, Zhang X, Li X, Hua Q, Dong L, et al. Flexible quantum dot-sensitized solar cells employing CoS nanorod arrays/graphite paper as effective counter electrodes. Journal of Materials Chemistry A. 2014;**2**(33):13661-13666. DOI: 10.1039/C4TA02052A

[64] Dang MT, Lefebvre J, Wuest JD. Recycling indium tin oxide (ITO) electrodes used in thin-film devices with adjacent hole-transport layers of metal oxides. ACS Sustainable Chemistry & Engineering. 2015;**3**(12):3373-3381. DOI: 10.1021/acssuschemeng.5b01080

[65] Guo Y, Zhong M, Fang Z, Wan P, Yu G. A wearable transient pressure sensor made with MXene Nanosheets for sensitive broad-range human-machine interfacing. Nano Letters. 2019;**19**(2):1143-1150

[66] Shi L, Meng L, Jiang F, Ge Y, Li F, Wu X, et al. In situ inkjet printing strategy for fabricating perovskite quantum dot patterns. Advanced Functional Materials. 2019;**29**(37):1903648. Available from: https://onlinelibrary.wiley.com/doi/abs/10.1002/adfm.201903648

[67] Cruz S, Dias D, Viana JC, Rocha LA. Inkjet printed pressure sensing platform for postural imbalance monitoring. IEEE Transactions on Instrumentation and Measurement. 2015;**64**(10):2813-2820

[68] Wang Y-F, Sekine T, Takeda Y, Yokosawa K, Matsui H, Kumaki D, et al. Fully printed PEDOT:PSS-based temperature sensor with high humidity stability for wireless healthcare monitoring. Scientific Reports. 2020;**10**(1):2467. DOI: 10.1038/s41598-020-59432-2

[69] Bihar E, Corzo D, Hidalgo TC, Rosas-Villalva D, Salama KN, Inal S, et al. Fully inkjet-printed, ultrathin and conformable organic photovoltaics as power source based on cross-linked PEDOT:PSS electrodes. Adv. Materials and Technologies. 2020;**5**(8):2000226. Available from: https://onlinelibrary.wiley.com/doi/abs/10.1002/admt.202000226

[70] Valentini L, Cardinali M, Grkovic M, Uskokovic PS, Alimenti F, Roselli L, et al. Flexible transistors exploiting P3HT on paper substrate and graphene oxide film as gate dielectric: Proof of concept. Science of Advanced Materials. 2013;**5**(5):530-533

[71] Feast WJ, Tsibouklis J, Pouwer KL, Groenendaal L, Meijer EW. Synthesis, processing and material properties of conjugated polymers. Polymer. 1996;**37**(22):5017-5047. Available from: http://www.sciencedirect.com/science/article/pii/0032386196004399

[72] Garlapati SK, Divya M, Breitung B, Kruk R, Hahn H, Dasgupta S. Printed electronics based on inorganic semiconductors: From processes and materials to devices. Advanced Materials. 2018;**30**(40):1707600. Available from:

https://onlinelibrary.wiley.com/doi/abs/10.1002/adma.201707600

[73] Chen C-B, Kao H-L, Chang L-C, Cho C-L, Lin Y-C, Huang C-C, et al. Fabrication of inkjet-printed carbon nanotube for enhanced mechanical and strain-sensing performance. ECS Journal of Solid State Science and Technology. 2021;**10**(12):121001. DOI: 10.1149/2162-8777/ac40d4

[74] Aziz A, Bazbouz MB, Welland ME. Double-walled carbon nanotubes ink for high-conductivity flexible electrodes. ACS Applied Nano Materials. 2020;**3**(9):9385-9392. DOI: 10.1021/acsanm.0c02013

[75] Akindoyo JO, Ismail NH, Mariatti M. Development of environmentally friendly inkjet printable carbon nanotube-based conductive ink for flexible sensors: Effects of concentration and functionalization. Journal of Materials Science: Materials in Electronics. 2021;**32**(9):12648-12660. DOI: 10.1007/s10854-021-05900-y

[76] Ervasti H, Järvinen T, Pitkänen O, Bozó É, Hiitola-Keinänen J, Huttunen O-H, et al. Inkjet-deposited Single-Wall carbon nanotube micropatterns on stretchable PDMS-Ag substrate–electrode structures for Piezoresistive strain sensing. ACS Applied Materials & Interfaces. 2021;**13**(23):27284-27294. DOI: 10.1021/acsami.1c04397

[77] Pandhi T, Chandnani A, Subbaraman H, Estrada D. A review of inkjet printed graphene and carbon nanotubes based gas sensors. Sensors. 2020;**20**(19):5642

[78] Saidina DS, Eawwiboonthanakit N, Mariatti M, Fontana S, Hérold C. Recent development of graphene-based ink and other conductive material-based inks for flexible electronics. Journal of Electronic Materials. 2019;**48**(6):3428-3450

[79] Huckaba AJ, Garcia-Benito I, Kanda H, Shibayama N, Oveisi E, Kinge S, et al. Inkjet-printed TiO2/fullerene composite films for planar perovskite solar cells. Helvetica Chimica Acta. 2020;**103**(5):e2000044. DOI: 10.1002/hlca.202000044

[80] Winarski DJ. Development of zinc oxide based flexible electronics. Bowling Green State University; 2019

[81] Mirza F, Sahasrabuddhe RR, Baptist JR, Wijesundara MBJ, Lee WH, Popa DO. Piezoresistive pressure sensor array for robotic skin. Popa D, MBJ W, editors. Sensors for Next-Generation Robotics III. Baltimore, Maryland, United States: SPIE; 2016. pp. 168-179

[82] Wang L, Peng H, Wang X, Chen X, Yang C, Yang B, et al. PDMS/MWCNT-based tactile sensor array with coplanar electrodes for crosstalk suppression. Microsystems & Nanoengineering. 2016;**2**(1):16065. DOI: 10.1038/micronano.2016.65

[83] Wang Y-F, Sekine T, Takeda Y, Yokosawa K, Matsui H, Kumaki D, et al. Fully printed PEDOT:PSS-based temperature sensor with high humidity stability for wireless healthcare monitoring. Scientific Reports. 2020;**10**:2467

[84] Ozioko O, Kumaresan Y, Dahiya R. Carbon nanotube/PEDOT: PSS composite-based flexible temperature sensor with enhanced response and recovery time. In: 2020 IEEE International Conference on Flexible and Printable Sensors and Systems (FLEPS); 2020. pp. 1-4

[85] Yin Z, Tian B, Zhu Q , Duan C. Characterization and application of

PVDF and its copolymer films prepared by spin-coating and langmuir-Blodgett method. Polymers. 2019;**11**(12):2033

[86] Haque RI, Vié R, Germainy M, Valbin L, Benaben P, Boddaert X. Inkjet printing of high molecular weight PVDF-TrFE for flexible electronics. Flexible and Printed Electronics. 2015;**1**(1):15001. DOI: 10.1088/2058-8585/1/1/015001

[87] Hu X, Ding Z, Fei L, Xiang Y, lin Y. Wearable piezoelectric nanogenerators based on reduced graphene oxide and in situ polarization-enhanced PVDF-TrFE films. Journal of Materials Science. 2019;**54**(8):6401-6409. DOI: 10.1007/s10853-019-03339-5

[88] Suganuma K. Introduction to Printed Electronics. 1st ed. New York: Springer New York; 2014. p. 124

[89] Riheen MA, Saha TK, Sekhar P. Inkjet printing on PET substrate. Journal of the Electrochemical Society. 2019;**166**:B3036-B3039

[90] Ball AK, Das R, Roy SS, Kisku DR, Murmu NC. Experimentation modelling and optimization of electrohydrodynamic inkjet microfabrication approach: A Taguchi regression analysis. Sadhana-Academy Proceedings in Engineering Sciences. 2019;**44**(7):1-16. DOI: 10.1007/s12046-019-1146-5

[91] Bucciarelli A, Adami A, Chandaiahgari CR, Lorenzelli L. Multivariable optimization of inkjet printing process of Ag nanoparticle ink on Kapton. In: 2020 IEEE International Conference on Flexible and Printable Sensors and Systems (FLEPS); 2020. pp. 1-4

[92] i Garcia ER. Inkjet printed microelectronic devices and circuits. Bellaterra, Catalonia: Universitat Autònoma de Barcelona; 2014

[93] Liu Y, Derby B. Experimental study of the parameters for stable drop-on-demand inkjet performance. Physics of Fluids. 2019;**31**(3):32004. DOI: 10.1063/1.5085868

[94] Delrot P, Modestino MA, Gallaire F, Psaltis D, Moser C. Inkjet printing of viscous monodisperse microdroplets by laser-induced flow focusing. Physical Review Applied. 2016;**6**(2):24003. Available from: https://link.aps.org/doi/10.1103/PhysRevApplied.6.024003

[95] Du Z, Yu X, Han Y. Inkjet printing of viscoelastic polymer inks. Chinese Chemical Letters. 2018;**29**(3):399-404. Available from: https://www.sciencedirect.com/science/article/pii/S100184171730373X

[96] Lee C, Kang BJ, Oh JH. High-resolution conductive patterns fabricated by inkjet printing and spin coating on wettability-controlled surfaces. Thin Solid Films. 2016;**616**:238-246. Available from: https://www.sciencedirect.com/science/article/pii/S0040609016304497

[97] Soltman D, Subramanian V. Inkjet-printed line morphologies and temperature control of the coffee ring effect. Langmuir. 2008;**24**(5):2224-2231. DOI: 10.1021/la7026847

[98] Trantidou T, Elani Y, Parsons E, Ces O. Hydrophilic surface modification of PDMS for droplet microfluidics using a simple, quick, and robust method via PVA deposition. Microsystems & Nanoengineering. 2017;**3**(1):16091. DOI: 10.1038/micronano.2016.91

[99] Eom JS, Kim SH. Plasma surface treatment of polyimide for adhesive Cu/80Ni20Cr/PI flexible copper clad laminate. Thin Solid Films. 2008;**516**(14):4530-4534. Available from: https://www.sciencedirect.com/science/article/pii/S0040609008001077

[100] Yang S-Y, Yang H-X, Hu A-J. Chapter 4- super engineering plastics and forms. Advanced Polyimide Materials. In: Yang S-Y, editor. Amsterdam, Netherlands: Elsevier; 2018. pp. 137-193

[101] Cruz S, Rocha L, Viana J. Enhanced printability of thermoplastic polyurethane substrates by silica particles surface interactions. Applied Surface Science. 2015;**360**:198-206

[102] Abu-Khalaf J, Al-Ghussain L, Nadi A, Saraireh R, Rabayah A, Altarazi S, et al. Optimization of geometry parameters of inkjet-printed silver nanoparticle traces on PDMS substrates using response surface methodology. Materials. 2019;**12**(20):3329

[103] Podhajny RM. Corona treatment of polymeric films. Journal of Plastic Film & Sheeting. 1988;**4**(3):177-188. DOI: 10.1177/875608798800400303

[104] Schlozer R. The top 4 drying methods for inkjet. Inkjet Insight. 2020. Available from: https://inkjetinsight.com/knowledge-base/the-top-4-drying-methods-for-inkjet/ [Internet]

[105] Li Y, Yang Q, Li M, Song Y. Rate-dependent interface capture beyond the coffee-ring effect. Scientific Reports. 2016;**6**(1):24628. DOI: 10.1038/srep24628

[106] Al-Milaji KN, Zhao H. New perspective of mitigating the coffee-ring effect: Interfacial assembly. Journal of Physical Chemistry C. 2019;**123**(19):12029-12041. DOI: 10.1021/acs.jpcc.9b00797

[107] Majumder M, Rendall CS, Eukel JA, Wang JYL, Behabtu N, Pint CL, et al. Overcoming the "coffee-stain" effect by compositional Marangoni-flow-assisted drop-drying. The Journal of Physical Chemistry. B. 2012;**116**(22):6536-6542. DOI: 10.1021/jp3009628

[108] Chuang MY. Inkjet printing of Ag nanoparticles using dimatix inkjet printer, No 2. technical report. United States: University of Pennsylvania; 2017

[109] Abbas A, Bajwa I. Inkjet printing of Ag nanoparticles using Dimatix inkjet printer, No 1. Technical Report. United States: University of Pennsylvania; 2017

[110] Mujal J, Ramon E, Diaz E, Carrabina J, Calleja Á, Martínez R, et al. Inkjet printed antennas for NFC systems. In: 2010 IEEE International Conference on Electronics, Circuits and Systems ICECS 2010- Proceeding; 2010. pp. 1220-1223

[111] Sowade E, Mitra KY, Ramon E, Martinez-Domingo C, Villani F, Loffredo F, et al. Up-scaling of the manufacturing of all-inkjet-printed organic thin-film transistors: Device performance and manufacturing yield of transistor arrays. Organic Electronics. 2016;**30**:237-246. Available from: http://www.sciencedirect.com/science/article/pii/S156611991530238X

[112] Ramon E, Sowade E, Martínez-Domingo C, Mitra KY, Alcalde A, Baumann RR, et al. Large-scale fabrication of all-inkjet-printed resistors and WORM memories on flexible polymer films with high yield and stability. Flexible and Printed Electronics. 2021;**6**(1):15003. DOI: 10.1088/2058-8585/abdb40

[113] Scattareggia S. Inkjet application process development using functional fluids. IARIGAI: The International association of research organizations for the information, media and graphic arts industries. Finland: Aalto University; 2015. p. 23

[114] Lindh M. Inkjet Deposition of Electrolyte: Towards Fully Printed Light-Emitting Electrochemical Cells. Umeå University; 2013

[115] Patnaik DN, Subramanian V, Salahuddin SS. Three-dimensional structure formation via inkjet-printed metal nanoparticles: Ink and application development. Berkeley: University of California; 2018

[116] Hamad AH, Salman MI, Mian A. Effect of driving waveform on size and velocity of generated droplets of nanosilver ink (Smartink). Manufacturing Letters. Elsevier. 2020;**24**:14-18

[117] Kwon K-S, Jang M-H, Park HY, Ko H-S. An inkjet vision measurement technique for high-frequency jetting. The Review of Scientific Instruments. 2014;**85**(6):65101. DOI: 10.1063/1.4879824

[118] Kwon J, Baek S, Lee Y, Tokito S, Jung S. Layout-to-bitmap conversion and design rules for inkjet-printed Large-scale integrated circuits. Langmuir. 2021;**37**(36):10692-10701. DOI: 10.1021/acs.langmuir.1c01296

[119] Vila F, Pallarès J, Ramon E, Terés L. A systematic study of pattern compensation methods for all-inkjet printing processes. IEEE Transactions on Components, Packaging and Manufacturing Technology. 2016;**6**(4):630-636

[120] Solar Cells Reimagined. [cited 2020 Sep 7]. Available from: https://sauletech.com/

[121] Connolly KB. Johnnie Walker 'Smart Bottle' Performs for Consumers and Supply Chain [Internet]. 2015. Available from: https://www.packagingdigest.com/beverage-packaging/johnnie-walker-smart-bottle-performs-for-consumers-and-supply-chain150306

[122] Jia H. Who will win the future of display technologies? National Science Review. 2018;**5**(3):427-431. DOI: 10.1093/nsr/nwy050

[123] Hengge M, Livanov K, Zamoshchik N, Hermerschmidt F, List-Kratochvil EJW. ITO-free OLEDs utilizing inkjet-printed and low temperature plasma-sintered Ag electrodes. Flexible and Printed Electronics. 2021;**6**(1):15009. DOI: 10.1088/2058-8585/abe604

[124] Jürgensen N, Pietsch M, Hai X, Schlisske S, Hernandez-Sosa G. Green ink formulation for inkjet printed transparent electrodes in OLEDs on biodegradable substrates. Synthetic Metals. 2021;**282**:116930. Available from: https://www.sciencedirect.com/science/article/pii/S0379677921002368

[125] Cinquino M, Prontera CT, Zizzari A, Giuri A, Pugliese M, Giannuzzi R, et al. Effect of surface tension and drying time on inkjet-printed PEDOT:PSS for ITO-free OLED devices. Journal of Science: Advanced Materials and Devices. 2022;**7**(1):100394. Available from: https://www.sciencedirect.com/science/article/pii/S2468217921000708

[126] Lee SY, Lee G, Kim DY, Jang SH, Choi I, Park J, et al. Investigation of high-performance perovskite nanocrystals for inkjet-printed color conversion layers with superior color purity. APL Photonics. 2021;**6**(5):56104. DOI: 10.1063/5.0044284

[127] Triana MA, Hsiang E-L, Zhang C, Dong Y, Wu S-T. Luminescent nanomaterials for energy-efficient display and healthcare. ACS Energy Letters. 2022;**7**:1001-1020

[128] Alamri AM, Leung S, Vaseem M, Shamim A, He J-H. Fully inkjet-printed photodetector using a graphene/perovskite/graphene Heterostructure. IEEE Transactions on Electron Devices. 2019;**66**(6):2657-2661

[129] Schackmar F, Eggers H, Frericks M, Richards BS, Lemmer U,

Hernandez-Sosa G, et al. Perovskite solar cells with all-inkjet-printed absorber and charge transport layers. Adv. Materials and Technologies. 2021;**6**(2):2000271. Available from: https://onlinelibrary.wiley.com/doi/abs/10.1002/admt.202000271

[130] Sajedi-Moghaddam A, Rahmanian E, Naseri N. Inkjet-printing Technology for Supercapacitor Application: Current state and perspectives. ACS Applied Materials & Interfaces. 2020;**12**(31):34487-34504. DOI: 10.1021/acsami.0c07689

[131] Sollami Delekta S, Laurila M-M, Mäntysalo M, Li J. Drying-mediated self-assembly of graphene for inkjet printing of high-rate Micro-supercapacitors. Nano-Micro Letters. 2020;**12**(1):40. DOI: 10.1007/s40820-020-0368-8

[132] Sollami Delekta S, Adolfsson KH, Benyahia Erdal N, Hakkarainen M, Östling M, Li J. Fully inkjet printed ultrathin microsupercapacitors based on graphene electrodes and a nano-graphene oxide electrolyte. Nanoscale. 2019;**11**(21):10172-10177. DOI: 10.1039/C9NR01427F

[133] Li Z, Ruiz V, Mishukova V, Wan Q, Liu H, Xue H, et al. Inkjet printed disposable high-rate on-paper Microsupercapacitors. Advanced Functional Materials. 2022;**32**(1):2108773. DOI: 10.1002/adfm.202108773

[134] Giannakou P, Tas MO, Le Borgne B, Shkunov M. Water-transferred, inkjet-printed Supercapacitors toward conformal and epidermal energy storage. ACS Applied Materials & Interfaces. 2020;**12**(7):8456-8465. DOI: 10.1021/acsami.9b21283

[135] Huang T-T, Wu W. Inkjet-printed wearable Nanosystems for self-powered technologies. Advanced Materials Interfaces. 2020;**7**(12):2000015. DOI: 10.1002/admi.202000015

[136] VartaAG. Printed Electronics for Medical and Lesiure Use. 2020 [cited 2021 Aug 24]. Available from: https://www.advancedbatteriesresearch.com/articles/20633/printed-electronics-for-medical-and-lesiure-use

[137] Gong S, Schwalb W, Wang Y, Chen Y, Tang Y, Si J, et al. A wearable and highly sensitive pressure sensor with ultrathin gold nanowires. Nature Communications. 2014;**5**:1-8

[138] Yang Z, Wang D-Y, Pang Y, Li Y-X, Wang Q, Zhang T-Y, et al. Simultaneously detecting subtle and intensive human motions based on a silver nanoparticles bridged graphene strain sensor. ACS Applied Materials & Interfaces. 2018;**10**(4):3948-3954. DOI: 10.1021/acsami.7b16284

[139] Pan L, Chortos A, Yu G, Wang Y, Isaacson S, Allen R, et al. An ultra-sensitive resistive pressure sensor based on hollow-sphere microstructure induced elasticity in conducting polymer film. Nature Communications. 2014;**5**(1):3002. DOI: 10.1038/ncomms4002

[140] Shi J, Li X, Cheng H, Liu Z, Zhao L, Yang T, et al. Graphene reinforced carbon nanotube networks for wearable strain sensors. Advanced Functional Materials. 2016;**26**(13):2078-2084. DOI: 10.1002/adfm.201504804

[141] Cardenas JA, Andrews JB, Noyce SG, Franklin AD. Carbon nanotube electronics for IoT sensors. Nano Futures. 2020;**4**(1):12001. DOI: 10.1088/2399-1984/ab5f20

[142] Mousavi S, Howard D, Zhang F, Leng J, Wang CH. Direct 3D printing of highly anisotropic, flexible, constriction-resistive sensors for multidirectional

proprioception in soft robots. ACS Applied Materials & Interfaces. 2020;**12**: 15631-15643

[143] Zare Y, Rhee KY. Simulation of percolation threshold, Tunneling distance, and conductivity for carbon nanotube (CNT)-reinforced nanocomposites assuming effective CNT concentration. Polymers. 2020;**12**(1):114. Available from: https://www.mdpi.com/2073-4360/12/1/114

[144] Chen Z, Wang Z, Li X, Lin Y, Luo N, Long M, et al. Flexible piezoelectric-induced pressure sensors for static measurements based on nanowires/graphene Heterostructures. ACS Nano. 2017;**11**(5):4507-4513. DOI: 10.1021/acsnano.6b08027

[145] Chen S, Wu N, Ma L, Lin S, Yuan F, Xu Z, et al. Noncontact heartbeat and respiration monitoring based on a hollow microstructured self-powered pressure sensor. ACS Applied Materials & Interfaces. 2018;**10**(4):3660-3667. DOI: 10.1021/acsami.7b17723

[146] Lipomi DJ, Vosgueritchian M, Tee BC-K, Hellstrom SL, Lee JA, Fox CH, et al. Skin-like pressure and strain sensors based on transparent elastic films of carbon nanotubes. Nature Nanotechnology. 2011;**6**(12):788-792. DOI: 10.1038/nnano.2011.184

[147] Farooqui MF, Shamim A. Low cost inkjet printed smart bandage for wireless monitoring of chronic wounds. Scientific Reports. 2016;**6**(1):28949. DOI: 10.1038/srep28949

[148] Kuzubasoglu BA, Sayar E, Bahadir SK. Inkjet-printed CNT/PEDOT:PSS temperature sensor on a textile substrate for wearable intelligent systems. IEEE Sensors Journal. 2021;**21**(12):13090-13097

[149] Lo L-W, Zhao J, Wan H, Wang Y, Chakrabartty S, Wang C. An inkjet-printed PEDOT:PSS-based stretchable conductor for wearable health monitoring device applications. ACS Applied Materials & Interfaces. 2021;**13**(18):21693-21702. DOI: 10.1021/acsami.1c00537

[150] Ryu G, You J, Kostianovskii V, Lee E, Kim Y, Park C, et al. Flexible and printed PPG sensors for estimation of drowsiness. IEEE Transactions on Electron Devices. 2018;**65**(7):2997-3004

[151] Gupta R. Biomedical sensors and their interfacing. Smart Sensors, Measurement and Instrumentation. 2017;**25**:219-248

[152] Welch KC, Kulkarni AS, Jimenez AM, Douglas B. Wearable sensing devices for human-machine interaction systems. In: 2018 United States National Committee of URSI National Radio Science Meeting (USNC-URSI NRSM); 2018. pp. 1-2

[153] Liao Y, Zhang R, Wang H, Ye S, Zhou Y, Ma T, et al. Highly conductive carbon-based aqueous inks toward electroluminescent devices{,} printed capacitive sensors and flexible wearable electronics. RSC Advances. 2019;**9**(27):15184-15189. DOI: 10.1039/C9RA01721F

[154] Mass M, Veiga LS, Garate O, Longinotti G, Moya A, Ramón E, et al. Fully inkjet-printed biosensors fabricated with a highly stable ink based on carbon nanotubes and enzyme-functionalized nanoparticles. Nanomaterials. 2021;**11**(7):1645

[155] Schultz A, Knoll T, Urban A, Schuck H, von Briesen H, Germann A, et al. Novel cost-efficient graphene-based impedance biosensor for the analysis of viral Cytopathogenicity and

the effect of antiviral drugs. Frontiers in Bioengineering and Biotechnology. 2021;**9**:718889. Available from: https://pubmed.ncbi.nlm.nih.gov/34381768

[156] Bihar E, Wustoni S, Pappa AM, Salama KN, Baran D, Inal S. A fully inkjet-printed disposable glucose sensor on paper. npj Flexible Electronics. 2018;**2**(1):30. DOI: 10.1038/s41528-018-0044-y

[157] Wang L, Wu Z, Cao C. Technologies and fabrication of intelligent packaging for perishable products. Applied Sciences. 2019;**9**(22):4858

[158] Urbano O, Perles A, Pedraza C, Rubio-Arraez S, Castelló ML, Ortola MD, et al. Cost-effective implementation of a temperature traceability system based on smart RFID tags and IoT services. Sensors. 2020;**20**(4):1163

[159] Zikulnig J, Hirschl C, Rauter L, Krivec M, Lammer H, Riemelmoser F, et al. Inkjet printing and characterisation of a resistive temperature sensor on paper substrate. Flexible and Printed Electronics. 2019;**4**(1):15008. DOI: 10.1088/2058-8585/ab0cea

[160] Quintero AV, Molina-Lopez F, Smits ECP, Danesh E, van den Brand J, Persaud K, et al. Smart RFID label with a printed multisensor platform for environmental monitoring. Flexible and Printed Electronics. 2016;**1**(2):25003. DOI: 10.1088/2058-8585/1/2/025003

[161] Baumbauer CL, Anderson MG, Ting J, Sreekumar A, Rabaey JM, Arias AC, et al. Printed, flexible, compact UHF-RFID sensor tags enabled by hybrid electronics. Scientific Reports. 2020;**10**(1):16543. DOI: 10.1038/s41598-020-73471-9

[162] Machiels J, Appeltans R, Bauer DK, Segers E, Henckens Z, Van Rompaey W, et al. Screen printed antennas on Fiber-based substrates for sustainable HF RFID assisted E-fulfilment smart packaging. Materials. 2021;**14**(19):5500

[163] Huang Q, Zhu Y. Printing conductive nanomaterials for flexible and stretchable electronics: A review of materials, processes, and applications. Advanced Materials Technologies. 2019;**4**(5):1-41

[164] Abdolmaleki H, Kidmose P, Agarwala S. Droplet-based techniques for printing of functional inks for flexible physical sensors. Advanced Materials. 2021;**33**(20):2006792. DOI: 10.1002/adma.202006792

[165] Wiklund J, Karakoç A, Palko T, Yiğitler H, Ruttik K, Jäntti R, et al. A review on printed electronics: Fabrication methods, inks, substrates, applications and environmental impacts. Journal of Manufacturing and Materials Processing. 2021;**5**(3):89

[166] Guo Y, Patanwala HS, Bognet B, Ma AWK. Inkjet and inkjet-based 3D printing: Connecting fluid properties and printing performance. Rapid Prototyping Journal. 2017;**23**(3):562-576. DOI: 10.1108/RPJ-05-2016-0076

Chapter 4

The Requirements of Product Lifecycle Management (PLM) Frameworks for Integration and Synergic Collaboration with Omnichannel Strategy

Noushin Mohammadian and Omid Fatahi Valilai

Abstract

The Importance of Product Lifecycle Management (PLM) is inevitable in fulfilling the collaboration and integration of different disciplines engaged in product development processes. However, PLM has faced with challenges in the context of Industry 4.0; especially in alignment with the Omnichannel strategy. The integration of the processes related to product promotion and digital marketing is recognized as one of those dominant challenges. There is an essential requirement for improving the interaction of PLM disciplines and Omnichannel processes. This chapter is investigating the challenges and restrictions raised by the Omnichannel strategy for PLM and the capabilities of the Omnichannel strategy to increase the effectiveness of PLM frameworks. Considering the Omnichannel strategy capabilities which are highly active in sales and promotions, the characteristics of bidirectional relation with PLM disciplines have been focused. For creating this bidirectional and synergic relation among Omnichannel strategy and PLM disciplines, the following perspectives are essential. Firstly, the configuration of Omnichannel strategy as an active agent among PLM disciplines. Secondly, the Omnichannel strategy's capabilities to solve challenges and problems in PLM. Moreover, the consequences of decisions and interactions when integrating Omnichannel strategy in the PLM framework will be worthy to be analyzed.

Keywords: Omnichannel, PLM, channel integration, industry 4.0

1. Introduction

Nowadays retail industries are changing sharply. Sales are not only happening in physical stores but also in many new touchpoints. One of the challenges is to coordinate these channels and have the same information among them e.g., price of the product, which refers to one of Omnichannel's features. More touchpoints mean more data; therefore, the need for data analytics is inevitable. With the help of data analytics tools, data that are generated from the touchpoints like social media can be analyzed and entered into each related PLM (Product Lifecycle Management) disciplines and make a change. The missing loop here is, data from PLM disciplines should also be transferred to the touchpoints because always all alterations cannot be practical and beneficial; therefore, before applying a change in the product lifecycle, the feedback whether positive or negative should as an input come to PLM, and the manufacturer should know the consequences of each small changes in design.

If channels work independently like the multi-channel concept, one of the issues would be creation of fragmented supply chains. Also, it would be difficult to deliver a consistent and reliable consumer experience. Besides, this independence leads to data mismatch, product/order information inconsistency, and poor inventory efficiency are more probable to happen. In the Omnichannel concept, the consumer can switch among channels easily e.g., they can search for a product in one channel and buy it in another channel and receive the product with a home delivery option which is the third channel.

Omnichannel has a vital role in sales by providing a seamless shopping experience for customers. The Omnichannel phase is more agile than before due to increasing the number of sales channels and also increasing the number and variety of new products. Customers are more demanding than before; therefore, to meet customer demands a strategy should be considered to be not only agile but also ahead of competitors. In the B2C phase, the most important factor which leads to profit is customer satisfaction. When a customer is satisfied, it is more plausible that another purchase is realized. Customer satisfaction is an important goal of Omnichannel strategy. For creating a seamless shopping experience different modes of delivery are considered in Omnichannel strategy like STS (ship-to-store), BOPS (buy-online-and-pick-up-in-store), Click & Collect, Click & Return, packet stations, home delivery, etc. Some of the benefits of applying Omnichannel are incremental sales due to new channels and devices, higher average sales to existing customers, a higher proportion of consumers making a purchase, promotional synergies across channels, lower inventory costs, reduced shipping costs due to in-store pick-up, and lower product returns.

The beneficial role of Omnichannel in sales and profit is clear, given that sales are also related to PLM and exist in PLM disciplines, and PLM has a vital role in all aspects and phases of the product, but a connection between Omnichannel and PLM is not considered yet. One of the objectives of this chapter is not only to relate these two concepts but also to consider a role for Omnichannel in PLM disciplines. Omnichannel is going to be embedded in the B2B concept.

For connecting PLM and Omnichannel a mediator is needed. Because Omnichannel is more agile and it grows rapidly day by day due to increasing the number of sales channels, increase the use of the internet, marketing strategies, etc. On the other hand, PLM is slower, and it is not like Omnichannel up to date. The relation between these two areas can happen through a mediator to balance the speed between them and also can transfer data.

2. Literature review

2.1 Evolution and definition of Omnichannel strategy

It started with the creation of "Single channel" which is categorized in two formats: B&M (Brick and Mortar) refers to the traditional physical stores in which demands are fulfilled at retail stores [1]. Direct channel refers to a channel which directly connected to the on-hand inventory from the top-level (It happens through the internet). Due to the direct connection between channel and manufacturer, inventory tracking is easy [2, 3].

Then "Dual-channel" or "Multi-channel" is introduced. It is defined as using more than one channel for sale products e.g., B&M and direct channel. Multi-channel comes with the benefits like products are more available for customers by using more than one channel, more profit in comparison with single-channel [4].

Closed-Loop-Supply-Chain (CLSC) is classified in the Multi-channel category which defines as after product is used by the customer, end-product is returned to the manufacturer for remanufacturing and re-distribution and keep cooperation of customers, they receive compensation [5].

Cross-channel conception and cross-channeling are synonymous terms in media marketing to designate the properties of products or services that are marketed via various media, telecommunications, or institutional channels [6]. Communication and sales take place through different channels that are interconnected. A customer can therefore obtain information across all channels. The networking of the channels is noticeable to the customer- behind it, there is a common database [7]. Even though cross-channel helps in many ways but there is still a problem: the system relies on the customer to act as an integrator of the data; Furthermore, in cross-channel strategy, there is a lack of centralized product knowledge base for the supply chain to synchronize whole channels. To reach this level of matureness, the omnichannel concept has been launched to have a holistic view of all channels for both consumer and supply chain members [8].

The first definition of omnichannel retailing is defined as "an integration of sales experience that melds the benefits of physical stores with the information and rich experience of online shopping" [9]. The term "Omni-retailing" for the first time was used in 2012 [10]. In 2013, the phrase 'Omni-retailing' as "a coordinated multichannel offering that provides a seamless experience when using all of the retailer's shopping channels" which indicates consumers use all channels and run into seamless integration is suggested [11]. In [12], is mentioned that the omnichannel idea emerged from the "click and mortar" context.

The concept of omnichannel is come from the evolution of multichannel, in multichannel, there is a gap between online and offline stores while in omnichannel customers can easily move among online stores, mobile devices, and B&M [13]. Even though multichannel retailing concentrates on the interaction among channels, omnichannel retailing goes to the upper level which is concentrated on the community and access to the mass/social media [14].

Omnichannel refers to the cooperative planning, management, and monitoring of the numerous available sales channels and customer touchpoints to optimize the customer experience and the company's success across the dissimilar sales channels and process steps. Customers can shift among the various channels (stationary, online, mobile, call center, social media, catalogs) at any time. Channels and brands interact with one another. Customer-contact points are the result of direct or indirect contact with a brand or company (together with retailers). Omnichannel management is sometimes referred to as cross-channel management. However, some see cross-channel as an intermediate phase in the integration of the multiple channels of the multi-channel on the way to becoming an Omnichannel. Although the sales channels are connected in the cross-channel, the procedure steps are not all professionalized. Full multi-channel integration will only be attained in the future due to legal, technological, organizational, and operational challenges [15]. The evolution of Omnichannel has been shown in **Figure 1**.

2.2 Challenges in Omnichannel strategy

The reasons which cause failure in implementing Omnichannel:

- Channels cannot communicate
- The real-time stock visibility in the e-commerce site cannot be guaranteed
- The staff has no way to know what's available where (have access to real-time inventory in all touchpoints)
- Cross-channel orders, returns, and exchanges cannot be offered
- Tracking and tracing items [16]
- Automation and centralization level of data management [16]
- Customer Experience in a complex multi-channel environment (provide a seamless shopping experience e.g. fast delivery, data about the availability of products in inventory, etc.) [16]
- Connecting the production and sales with the logistics and delivery services [16]

For implementing a successful Omnichannel strategy, the above-mentioned items should take into consideration. Based on the researches [17], the value of

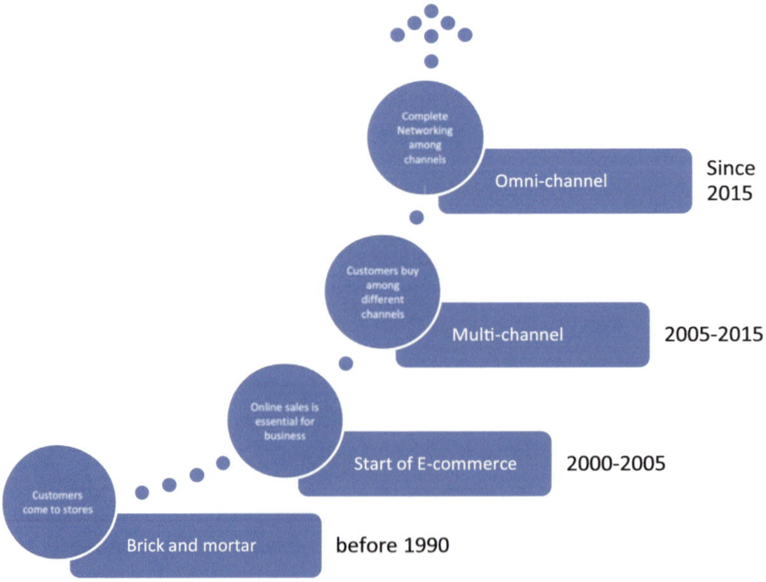

Figure 1.
Omnichannel evolution.

omnichannel could appear to increase customer loyalty, business revenue, and business agility.

2.3 Customer touchpoints

Customer touchpoints have many vital effects, e.g. between producers and customers, customers can get information about product/service and make appraisals at various points and have an interaction with retailers through several various channels across a complicated purchase procedure [18].

Secondly, it is playing an important role in attracting customers and it is a direct and indirect intermediary between customer and brand [19].

Touchpoints are [14, 19–21]:

- Traditional advertising media
- E-mail
- Catalogs
- Website
- Social media
- Brick and mortar
- Return
- TV
- Customer Service
- Payment
- Feedback
- Search Engines
- Phone calls
- Online reviews

Due to the importance of touchpoints, ways should be introduced to track (including real-time tracking, discrete vs. relational relationships) and have an effect on them. Touchpoints are where brands and customers interact. Therefore; it is one of the places to enhance engagement and profitability [22]. The results of touchpoints are gathered from pre-purchase, purchase, and post-purchase situations [23].

2.4 Integration in channels

As more channels are initiated, more data are gathered in each touchpoint quickly. These data should be collected and be analyzed. This leads to understanding the importance of data integration among channels, ignoring this integration, leads to customer dissatisfaction and enterprise will be faced with an increase in costs [24]; Furthermore, consumers hope for a uniform, consistent, integrated platform which by this, can seamlessly move through channels (physical store, online and mobile) [25], without seeing any contradictions for example in terms of pricing or availability of the product and this uniformity will generate positive experiences. When channels are working independently from each other, it causes fragmented supply chains and it makes lots of effort to transport a consistent and reliable customer experience [26].

For implementing integration in omnichannel a three dimensions framework can be taken into consideration:

1. Integration among channel stages helps to provide a smooth customer purchase experience, with the total awareness of product/service provider about the total journey of customer shopping in earlier stages and also the post-shopping stages. This integration can help the consumer to move through channels forward and backward without any perplexity, losing control of order, facing mismatched data about the product/service [27].

2. Integration among channel types makes synchronization between operations and decisions among various channel types (online, offline), and meanwhile, proper communication should take place among them. This integration makes it possible for the customer to change channel types [28].

3. Integration among channel agents: This integration makes sure that various channel agents send identical data and supply the same product/service to the customer [29].

These points show that the omnichannel system should access information all over the touchpoints and can notice any change in any channel in real-time and inform the whole system e.g. if a customer cancels the order, it should be detected and send command to different sections for example in terms of packaging, logistics, payment, etc. to apply the change and re-schedule the plan instantly.

2.5 Solutions for Omnichannel challenges with the help of industry 4.0

As mentioned before, there are challenges in Omnichannel strategy that can be solved with the help of technological developments like Industry 4.0 [30]. Based on Literature [31], Industry 4.0 is defined as "a collective term for technologies and concepts of value chain organization". IoT, Cyber-physical systems, the Internet of services, smart factories are defined as the core of Industry 4.0.

IoT helps in different ways to solve challenges in Omnichannel strategy for example in terms of considering IoT-in-store-environment using RFID, NFC, Beacon to track products and in this regard can prevent theft and provide real-time inventory field [30, 32–34]. IoT by using sensors can detect customers entering the shop (with their mobile devices) and identify them and with the connection to the Omnichannel's touchpoints can figure out the customer's shopping history or recently searched items and offer related products in-store to the customers. By detecting entering customers to the shop, the number of cashiers and staffs which can fulfill their needs can be easily determined and it helps by decreasing waiting time, increase customer satisfaction and provide a good experience for customers and finally the chance of re-visiting the store will be increased [35–38].

2.6 Product lifecycle management

2.6.1 Definition

The concept of PLM was initiated to describe the evolution of the product which is categorized as, promotion, maturity, and decline phases [39]. In the 1980s, by introducing concurrent engineering, gradually the concept of PLM is entered in the manufacturing engineering field and it causes new categories in PLM processes for example market analysis, product design, process development, product manufacturing, production distribution, product use, post-sale service and recycling [40].

Product lifecycle management is one of the concepts whose focus is on managing whole product data (live data) from the first step (ideation) until the last step (disposal) together with support the local data and information integration in the stage of product design and development, for instance: assisting designer to access information generated from CAD, CAM, CAE, etc.

2.6.2 Phases

1. Conceive [specification, concept design]

2. Design [detailed design, validation, and analysis (simulation), tool design]

3. Realize [plan manufacturing, manufacture, build/assemble, test]

4. Service [sell and deliver, use, maintain and support, dispose]

2.6.3 Stages

Product lifecycle consists of three stages [41]:

1. BOL (beginning of life) includes design and manufacturing

2. MOL (middle of life) includes product usage, service, and maintenance

3. EOL (end of life) includes disassembly, remanufacture, recycling, reuse, disposal

2.6.4 Techniques in PLM

These are some techniques used in PLM as shown in **Figure 2**:

Concurrent engineering workflow, Industrial design, Bottom-up design, Top-down design, Both-ends-against-the-middle design, Design in context, Modular design, NPD (new product development), DFSS (design for six sigma), Digital simulation engineering, Configuration management.

With the help of the internet and IT, efforts have been made to reduce costs, accelerate product development time and improve quality which are the main factors affecting

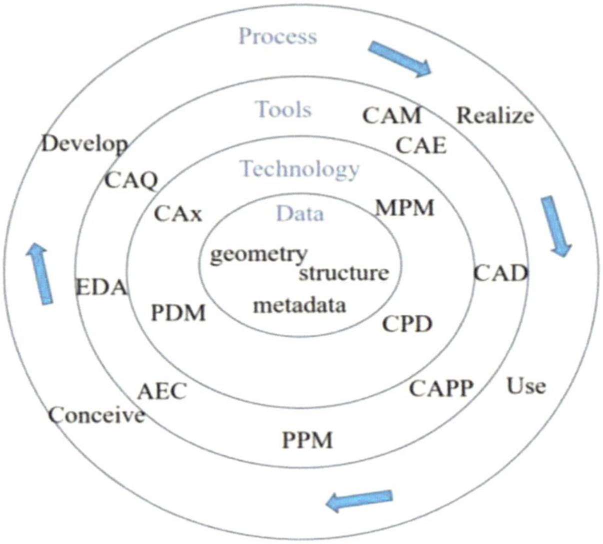

Figure 2.
Product lifecycle management [42].

customer satisfaction and in the same way increase profit, this is the main goal of the omnichannel strategy [43, 44]. PLM is the concept, which covers stages before production and according to omnichannel which has a great focus on post-production stages, the combination of these two concepts can help to develop omnichannel in the first stages of producing a product. The application of Big data in PLM outstandingly falls behind other areas, mainly for electronic commerce. The worse issue is: it becomes usual that manufacturers do not use stored data and they do not know how to use them [45].

2.6.5 New expectations of PLM

1. Shorten product lifecycle: due to the agile market, for being able to fulfill the needs of customers on time, it is important to decrease the time spent especially in the design and manufacturing phase.

2. Production based on modularity: Nowadays, customization and personalization have received enormous attention among customers; therefore, production should consider modular products to be more flexible and provides more responsive design to fulfill needs.

3. Automation and intelligent equipment: There is a competitive atmosphere between retailers to produce low-cost goods but in the same or even better quality and provide short-delivery-time, to satisfy these factors, manufacturing activities should lead to automation.

4. Remote services: Nowadays customers looking forward to receiving higher quality which makes the company, develop data monitoring, tracking, and also management system to be able to be more responsive in the online platform and solve problems, for instance: remote product repair, maintenance and upgrade.

Then as the above-mentioned factors are discussed, the focus of PLM is changing from product gradually to service and also from informatization to intellectualization, in this case applying AI is a very effective [40, 46].

2.6.6 The role of social media and its impact on PLM

The power of social media cannot be neglected in recent years. With the help of social media, it is possible to reach features for example in terms of rating a product, promoting a product/service, asking a question from others about the product, sharing opinions about the product in real-time, sharing videos, and pictures about products (when it is positively, acts as an advertising and it will be caused more sales), show satisfaction/dissatisfaction about product/service.

Employing customers as brand advocates, involving them at different phases of product design, and use of their capabilities to obtain focus groups to try new products/services and share their opinions on social media. Nowadays, social media is used as an extra sales channel and also has a big role in attracting and in some cases repelling customers, that is the reason why social media is important in the marketing [25]. By analyzing the information on social media, it is possible to implement these data in PLM phases, for example: the design phase.

2.6.7 Challenges in PLM

In some PLM solutions failure is noticed, the reason for this is complexity in integration. These concepts are not considered completely in PLM [47]:

1. major enterprises solutions for instance: CRM, ERP, logistic based Supply Chain Management

2. Marketing and sales

3. Distribution

4. HRM (Human resource management)

5. Finance

There is not an efficacious mechanism for knowledge and service exchange and sharing among the stakeholders in the product lifecycle [43].

2.6.8 Risks in implementing PLM

Based on information and communication technology (ICT), different kinds of risks are observed. These risks, in some cases, are related to technology, while in other cases they are related to the data and business processes, and also human-related risks are noticed, knowing them can help to reduce the probability of PLM adoption failure.

Some affected risk factors are listed below, these items should take into account during implementation [48]:

- Cultural changes

- Improper data migration
- Data security
- User acceptance
- Technology-process alignment
- Technology-business alignment

2.6.9 Impacts of industry 4.0 on PLM

PLM is considered as a solution to develop the product and helps to win in a competitive market [49]. Improvement in technologies helps companies to increase their productivity for example Industry 4.0 helps in terms of having a full data integration environment and automated processes related to the product [50, 51]. PLM is now busy with product data and processes, with the advancement of technology PLM should be able to manage and develop smart products and services. One of the important points is the ability to have a communicate in real-time and exchange data. With the help of Industry 4.0., all data through the whole product lifecycle can be collected and stored. As a result, Industry 4.0 developments provide new possibilities to obtain the entire lifecycle incorporation [52].

Maintenance is one of the disciplines in PLM which plays an important role in brand image, customer satisfaction, and profitability. Maintenance is a complicated phase that has a great effect on operating costs, and it is hard to manage. By using Industry 4.0., this phase can be managed easier and remotely (maintenance as a service) by having a real-time data [53].

As a result, Industry 4.0 has many potentials which can be used in PLM to make improvements, decrease costs, decrease production time, decrease waiting time (for customers), increase profitability for example in terms of collecting data in real-time, analyzing data for product design, product monitoring, product maintenance.

3. Perspectives of integrating product lifecycle management and Omnichannel strategy

Omnichannel strategy is a new concept with a sales-oriented attitude by increasing sales channels, with increasing use of the internet and marketing strategies that affect customers mentally and emotionally by analyzing data attracting more attention.

Based on Literature, Omnichannel strategy is working especially on sales, which is one of the disciplines in PLM. From the other sideshow studies, outcomes of data analytics in Omnichannel strategy are usually promoted to be used in PLM, however, the mutual interactions and relations of these concepts together are not observed yet [40, 54–57].

PLM has a great effect on the product, from the other side Omnichannel strategy has also affected the product, but these two concepts are not completely connected. The reason why there should be a connection between PLM and channels of Omnichannel is the important data that is generated in those channels. Before IoT, there wasn't so much data to create added value. The data of the channel's side (Omnichannel's channel) is related to customers and competitors; therefore, it is valuable. Another reason for this connection is that the Omnichannel strategy can solve problems in PLM disciplines.

For developing a framework for merging Omnichannel strategy and PLM literature, these points should be considered.

To deal with the topic we have started from four steps:

1. The mechanism for embedding Omnichannel strategy in the sales part of PLM

2. Find the category of problems in different disciplines of PLM which can be solved by Omnichannel strategy

3. Create a bi-directional framework between Omnichannel strategy and PLM

4. The way of solving problems in PLM with the help of Omnichannel strategy

3.1 The mechanism for embedding Omnichannel strategy in the sales through PLM

As mentioned before, Omnichannel strategy is working mainly on sales, and on the other hand, there are sales in one of the PLM disciplines; therefore, if Omnichannel strategy as an active agent be placed in the PLM sales department and be engaged from the first stage of producing a product it will come with improvements for example in terms of saving time, in this case in the first place, the way of selling, advertising, the place of selling (online, offline) and target group be considered from the beginning.

3.2 Find the category of problems in different disciplines of PLM which can be solved by Omnichannel strategy

The omnichannel strategy has many potentials which PLM can get the benefit of them to solve complications that happen among disciplines. For example, if a company has a problem selling products in a physical store, an Omnichannel strategy can help in terms of introducing other online channels. Or even it can reach new customers with the help of online services this can help increase sales and as a result, reduce costs for generating new channels [58, 59]. In the first place, the category of problems which Omnichannel strategy can find a solution for them should be prepared.

3.3 Create a bi-directional framework between Omnichannel strategy and PLM

Currently, there is a trend in which with the help of data analytics, data from the Omnichannel side (channels) is transferred to PLM and creates value. Just in the one-directional way. It means that data Analytics with analyzing data from the PLM side does not make a change on the Omnichannel side. If always one-side relations (from Omnichannel to PLM) are analyzed the result will be almost in a positive way but this is not lead to success every time, because the negative effects are neglected which can be realized through bi-directional relation it means that from omnichannel to PLM and also from PLM to Omnichannel side.

3.4 The way of solving problems in PLM with the help of Omnichannel strategy

After finding a category of problems that can be solved by Omnichannel strategy, it is time to figure out for each category of problems, which potentials of Omnichannel strategy can be matched. Instead of solving problems inside PLM disciplines through traditional ways which take more time and cost more for business, Omnichannel can recommend solutions that are more cost and time saving and therefore prevent lost sales and delays which lead to dissatisfaction and negatively affects profitability.

3.5 Sample loops of integrating Omnichannel strategy and PLM

This study aims to increase profit, customer satisfaction, convenience, see the cause and effect of modifications on customers before applying them, make use of Omnichannel whole abilities in PLM disciplines, investigate which problems can be solved by Omnichannel strategy and what is the effect of it, solve PLM's problems in Omnichannel environment. Some of the elements which are affected by both PLM and Omnichannel are discussed below, and cause and effect loops are explained.

1. Sales: sales are one of the important factors due to having a direct effect on profit which is one of the important factors for enterprises. Omnichannel strategy is sales-oriented, and in PLM, sales are defined in the third stage of it. If sales increase under the condition of good quality, positive reviews from the customers will be increased and then these reviews act as a marketing agent and result in

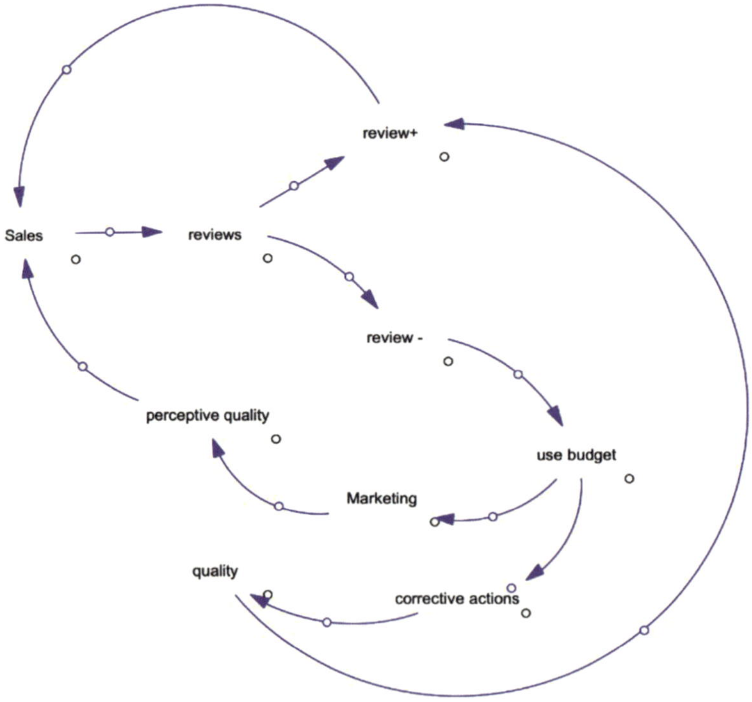

Figure 3.
A sample of the cause-and-effect loop of sales in Omnichannel strategy and PLM.

increasing the sales. In case of bad quality, two options can be considered: 1. refer to PLM and increase the quality by purchasing higher quality raw material, the problem here is, it takes time, and this delay causes lost sales then besides the lost sales, the budget is required for purchasing raw materials. 2. The suggestion can be as shown in **Figure 3** to use less budget in comparison with the first solution to increase perceptive quality instead of actual quality by using the potential of Omnichannel strategy through generating a positive trend (using an influencer in social media for advertising the product) about the product and create a positive mindset about the product which leads to increase sales but in the same time, the first solution should take into consideration because increasing perceptive quality is a fast solution which works for a short time but it helps to prevent lost sales and buy more time to deal with the quality. For applying this, the Omnichannel strategy should be embedded among PLM disciplines to be able to face problems on time and recommend a solution.

2. Price: the price is the most important decision factor for purchasing a product or using a service. It is set by the PLM side and has a great impact on customers which is related to the Omnichannel side. Price has a range, which is influenced by costs coming from the PLM side like quality, raw material; therefore, it cannot be placed under lower bound in Omnichannel side just for attracting more customers. If the price from Omnichannel side is decided to approach to lower bound, in PLM side some changes should take place like finding cheaper suppliers and cheaper suppliers decrease reputation then it results in decreasing the demand more.

In this case, the cost of production will be increased (because of lower demand) then the price will be set more than before and it results in decreasing demand again and therefore decreasing profit and the cost will be increased again, it will be a reinforcing loop as shown in **Figure 4**. For balancing, the cost of production should be decreased for example by selecting a cheaper supplier. For keeping the brand reputation, by using the potential of Omnichannel, through setting the mindset that the selected supplier has still quality (even though for example: if the supplier is recognized as a low-quality brand image).

3. Design: this item should be mostly based on customer needs and should meet the functionality of the product. From the Omnichannel side, data should be collected and after analyzing entered into PLM (especially design phase) phases. The problem of re-designing in PLM is, with considering customer needs modifications on design are applied and then expectations of customers will be increased and more re-designing is needed, and it is a negative reinforcing loop. With the help of Omnichannel expectations of customers can be changed or replaced with companies' expectations as shown in **Figure 5**. It can be aligned then with decreasing expectations the former loop will be balanced.

4. Quality: quality is essential because it has a direct effect on customer satisfaction which is one of the most important factors of Omnichannel strategy and it is determined by the PLM part. If the quality is lower than the desired level, more defects are expected, and more maintenance are applied then cost will be increased, and the result will be decreasing the profit. For decreasing maintenance costs, some costs should be reduced. For example, instead of normal workers using refugees

with lower income, but for justification by using Omnichannel strategy, a wave-based on "this company is using refugees to engage them more in the society"in this case, if there would a problem in maintenance it can be neglected by the consumer because they know the story behind it. By using the Omnichannel strategy, the cost will be decreased then the profit will be increased, and it will be used to enhance the quality as shown in **Figure 6**. Without considering the Omnichannel loop, occurring complications from maintenance workers, lead to dissatisfaction, and it affects brand image then results in decreasing sales.

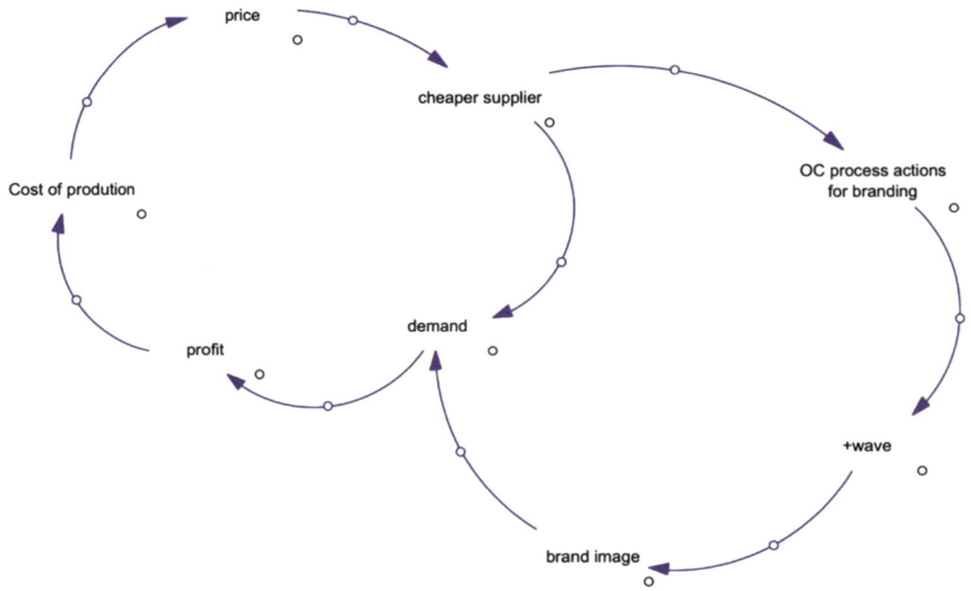

Figure 4.
A sample of the cause-and-effect loop for Price in PLM and Omnichannel strategy point of view.

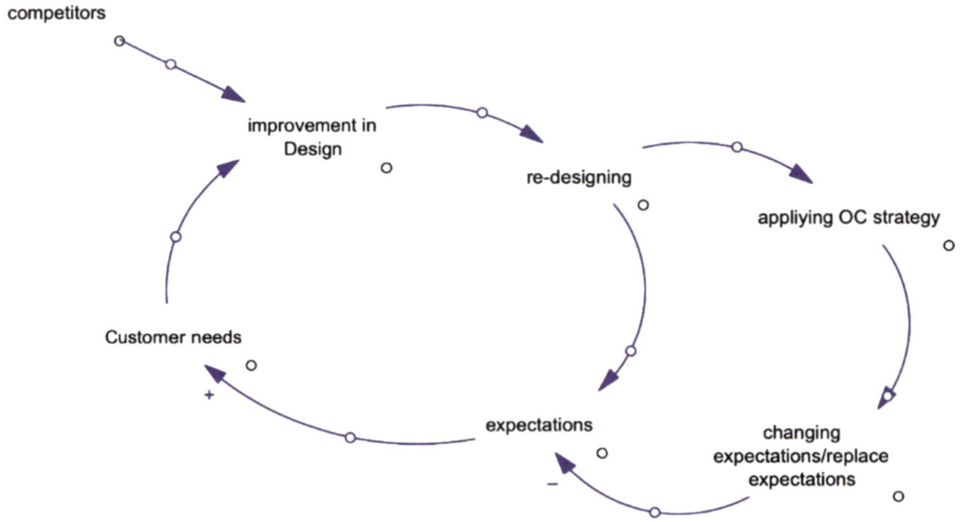

Figure 5.
A sample of the cause-and-effect loop of design from PLM and Omnichannel strategy point of view.

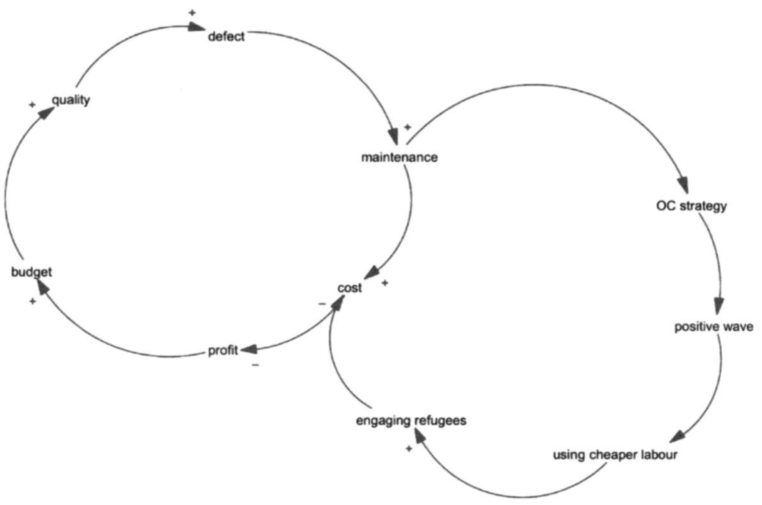

Figure 6.
A sample of the cause-and-effect loop of quality in perspective of PLM and Omnichannel strategy.

4. Conclusions

The main reason for the recent attention to customer satisfaction is competitiveness between companies, which leads to more concentration on the performance of processes for product/service's value which last in customer satisfaction. If customer satisfaction is not fulfilled, it makes lost sales.

One of the ways to reach customer satisfaction is customer experience, which is extremely important because, it starts a long time earlier than the time of shopping and it is a plus point to gain in competitive market [60], therefore; if this journey goes well it leads to purchase and one of customer satisfaction phase will be accomplished.

The omnichannel strategy should be integrated into three different stages: physical, business and application. At the first level, stock-keeping points and flows among channels should be connected and harmonized. At the second level, different channel processes and business models should cooperate. At the third level, information flows and databases should be in contact correctly [61]. With the increasing variety of products and increasing channels, the generation of data getting more than before; therefore, it is important to analyze data and extract vital information (customer's behavior, feedback, customer's needs, about competitors, etc.) to use in related phases to make improvement and increase profit. Omnichannel strategy is also playing an important role in marketing and can make a big difference in profit for producers, by implementing Omnichannel strategy in a brand, more customers are attracted to the brand, and it leads to more profit, but this strategy has many potentials which are not used yet, some aspects of it are discussed in this chapter. For instance, if there is an issue in the production part instead of changing whole disciplines (e.g., in the design phase, manufacturing, etc.), the problem can be solved through Omnichannel channels, by changing the idea of customers by inducing the band's idea with smart advertisement.

Author details

Noushin Mohammadian and Omid Fatahi Valilai*
Jacobs University, Bremen, Germany

*Address all correspondence to: o.fatahivalilai@jacobs-university.de

References

[1] Swami S, Shah J. Channel coordination in green supply chain management. The Journal of the Operational Research Society. 2013;**64**(3):336-351. DOI: 10.1057/jors.2012.44

[2] Rahman Z. Internet-based supply chain management: Using the internet to revolutionize your business. International Journal of Information Management. 2003;**23**:493-505

[3] Tetteh A. Supply chain distribution networks: Single-, dual- and Omni-channel. Interdisciplinary Journal of Research in Business. 2014;**3**:63-73

[4] Kevin Chiang W, Monahan GE. Managing inventories in a two-echelon dual-channel supply chain. European Journal of Operational Research. 2005;**162**(2):325-341

[5] Huang M, Song M, Lee LH, Ching WK. Analysis for the strategy of closed-loop supply chain with dual recycling channel. International Journal of Production Economics. 2013;**144**(2):510-520

[6] Shankar V, Kushwaha T. Omnichannel marketing: Are cross-channel effects symmetric? International Journal of Research in Marketing [Internet]. 2021;**38**(2):290-310. DOI: 10.1016/j.ijresmar.2020.09.001

[7] Alonso J, Federico G, Estela Nunez-Barriopedro P-M. Omnichannel management in B2B. Complexity-based model. Empirical evidence from a panel of experts based on fuzzy cognitive maps. Industrial Marketing Management. 2021;**95**:99-113

[8] Saghiri S, Wilding R, Mena C, Bourlakis M. Toward a three-dimensional framework for omni-channel. Journal of Business Research [Internet]. 2017;**77**(June 2016):53-67. DOI: 10.1016/j.jbusres.2017.03.025

[9] Rodriguez M. The future of shopping. Planning. 2018;**84**(9):10

[10] Bodhani A. Shops offer the e-tail experience. Engineering and Technology. 2012;**7**(5):46-49

[11] Levy M, Weitz BA, Grewal Professor D. Retailing Management. 9th ed. Vol. 2013. New York: McGraw-Hill Education; 2013

[12] Huang W. Literature review on Omnichannel retailing. Expert Journal of Marketing. 2021;**9**(1):1-7

[13] Rosenblum P, Kilcourse B. Omnichannel 2013: The long road to adaption. In: Benchmark Report. Miami: RSR Research; 2013

[14] Verhoef PC, Kannan PK, Inman JJ. From Multi-Channel retailing to Omni-Channel retailing. Introduction to the special issue on Multi-Channel retailing. Journal of Retailing. 2015;**91**(2):174-181. DOI: 10.1016/j.jretai.2015.02.005

[15] Oeser PDG. No Title. In

[16] Mirzabeiki V, Saghiri SS. From ambition to action: How to achieve integration in omni-channel? Journal of Business Research. 2020;**110**:1-11. DOI: 10.1016/j.jbusres.2019.12.028

[17] Fay DL. "Sistem Aplikasi Pesan Antar Makanan Berbasis web Di Omah Pawon" Angew. Angewandte Chemie International Edition. 1967;**6**(11):951-952

[18] Dorman AJ. Omni-Channel retail and the new age consumer: An empirical analysis of direct-to-Consumer Channel interaction in the retail industry. Chapter 2. CMC Senior Theses. 2013:1-38

[19] Baxendale S, Macdonald EK, Wilson HN. The impact of different touchpoints on brand consideration. Journal of Retailing. 2015;**91**(2):235-253

[20] Kushwaha T, Shankar V. More valuable? The moderating role. Journal of Marketing. 2013;77:67-85

[21] Verhoef PC, Neslin SA, Vroomen B. Multichannel customer management: Understanding the research-shopper phenomenon. International Journal of Research in Marketing. 2007;**24**(2):129-148

[22] Manser Payne E, Peltier JW, Barger VA. Omni-channel marketing, integrated marketing communications and consumer engagement. Journal of Research in Interactive Marketing. 2017;**11**(2):185-197

[23] Lemon KN, Verhoef PC. Understanding customer experience throughout the customer journey. Journal of Marketing. 2016;**80**(6):69-96

[24] Taufique Hossain TM, Akter S, Kattiyapornpong U, Wamba SF. The impact of integration quality on customer equity in data driven Omnichannel services marketing. Procedia Computer Science. 2017;**121**:784-790. DOI: 10.1016/j.procs.2017.11.101

[25] Piotrowicz W, Cuthbertson R. Introduction to the special issue information technology in retail: Toward omnichannel retailing. International Journal of Electronic Commerce. 2014;**18**(4):5-16

[26] Saghiri S, Wilding R, Mena C, Bourlakis M. Toward a three-dimensional framework for omni-channel. Journal of Business Research. 2017;77:53-67. DOI: 10.1016/j.jbusres.2017.03.025. DOI: 10.1007/978-3-319-31232-3_43

[27] Emrich O, Paul M, Rudolph T. Shopping benefits of multichannel assortment integration and the moderating role of retailer type. Journal of Retailing. 2015;**91**(2):326-342. DOI: 10.1016/j.jretai.2014.12.003

[28] Neslin SA, Shankar V. Key issues in multichannel customer management: Current knowledge and future directions. Journal of Interactive Marketing. 2009;**23**(1):70-81. DOI: 10.1016/j.intmar.2008.10.005

[29] Herhausen D, Binder J, Schoegel M, Herrmann A. Integrating bricks with clicks: Retailer-level and channel-level outcomes of online-Offline Channel integration. Journal of Retailing. 2015;**91**(2):309-325. DOI: 10.1016/j.jretai.2014.12.009

[30] Janhofer D, Barann B, Cordes AK, Becker J. Mastering omni-channel retailing challenges with industry 4.0 concepts. Vols. 2020-Janua. In: Proceedings of the Annual Hawaii International Conference on System Sciences. 2020. pp. 5389-5398

[31] Hermann M, Pentek T, Otto B, Pentek T. Design principles for industries 4.0 scenarios: A literature review competence Center corporate data quality (CC CDQ) view project industrial data space view project design principles for Industrie 4.0 scenarios: A literature review. In: 49th Hawaii Int Conf Syst Sci. Vol. 16. 2016

[32] Bandyopadhyay D, Sen J. Internet of things: Applications and challenges in technology and standardization. Wireless Personal Communications. 2011;**58**(1):49-69

[33] Burmester M, Munilla J, Ortiz A, Caballero-Gil P. An RFID-based smart structure for the supply chain: Resilient scanning proofs and ownership transfer with positive secrecy capacity channels. Sensors. 2017;**17**:1562

[34] Ping L, Liu Q, Zhou Z, Wang H. Agile supply chain management over the internet of things. International Conference on Management and Service Science. 2011;**2011**:16-19

[35] Lazaris C, Vrechopoulos A, Doukidis G, Fraidaki K. Mobile apps for Omnichannel retailing: Revealing the emerging showrooming phenomenon. In: MCIS. 2015

[36] Demirkan H, Spohrer J. Developing a framework to improve virtual shopping in digital malls with intelligent self-service systems. Journal of Retailing and Consumer Services. 2014;**21**(5):860-868

[37] Wahlster W, Kröner A, Schneider M, Baus J. Sharing memories of smart products and their consumers in instrumented environments. Information Technology. 2008;**50**(1):45-50

[38] Leukert B, Gläß R. Herausforderungen und Chancen für die Digitalisierung von Handelsunternehmen. In: Gläß R, Leukert B, editors. BT - Handel 4.0: Die Digitalisierung Des Handels – Strategien, Technologien, Transformation. Berlin, Heidelberg: Springer; 2017. pp. 193-211

[39] Levitt T. EXPLOIT the Product Life Cycle. Harv Bus Rev [Internet]. 1965;**43**(6):81-94. Available from: http://search.ebscohost.com/login.aspx?direct=true&db=bth&AN=3866434&lang=es&site=ehost-live

[40] Wang L, Liu Z, Liu A, Tao F. Artificial intelligence in product lifecycle management. International Journal of Advanced Manufacturing Technology. 2021;**114**(3-4):771-796

[41] Terzi S, Bouras A, Dutta D, Garetti M, Kiritsis D. Product lifecycle management - from its history to its new role. International Journal of Product Lifecycle Management. 2010;**4**(4):360-389

[42] Liu XL, Wang WM, Guo H, Barenji AV, Li Z, Huang GQ. Industrial blockchain based framework for product lifecycle management in industry 4.0. Robot Comput Integr Manuf [Internet]. 2020;**63**(January 2019):101897. DOI: 10.1016/j.rcim.2019.101897

[43] Liu XL, Wang WM, Guo H, Barenji AV, Li Z, Huang GQ. Industrial blockchain based framework for product lifecycle management in industry 4.0. Robotics and Computer-Integrated Manufacturing. 2020;**63**:101897. DOI: 10.1016/j.rcim.2019.101897

[44] Sudarsan R, Fenves SJ, Sriram RD, Wang F. A product information modeling framework for product lifecycle management. Computer-Aided Design. 2005;**37**(13):1399-1411

[45] Crawford K. BigData-SubmissionA - Six_Provocations_for_Big_Data.pdf. 2011:1-17. Available from: http://softwarestudies.com/cultural_analytics/Six_Provocations_for_Big_Data.pdf

[46] Tao F, Zhang L, Liu Y, Cheng Y, Wang L, Xu X. Manufacturing Service Management in Cloud Manufacturing: Overview and future research directions. Journal of Manufacturing Science and Engineering, Transactions of the ASME. 2015;**137**(4):040912-1

[47] Osman N, Sahraoui A-E-K. From PLM to ERP: A software systems engineering integration. International

Journal of Software Engineering & Applications. 2018;**9**(1):11-27

[48] Singh S, Chandra Misra S, Kumar S. Identification and ranking of the risk factors involved in PLM implementation. International Journal of Production Economics. 2020;**222**:107496. DOI: 10.1016/j.ijpe.2019.09.017

[49] d'Avolio E, Bandinelli R, Rinaldi R. Improving new product development in the fashion industry through product lifecycle management: A descriptive analysis. International Journal of Technology and Design Education. 2015;**8**(2):108-121

[50] Wang L, Törngren M, Onori M. Current status and advancement of cyber-physical systems in manufacturing. Journal of Manufacturing Systems. 2015;**37**:517-527

[51] Xu X. Robotics and computer-integrated manufacturing from cloud computing to cloud manufacturing ubiquitous product life cycle support. Robotics and Computer-Integrated Manufacturing. 2012;**28**(1):75-86

[52] Ferreira F, Marques AL, Faria J, Azevedo A. Large Project Management in the automotive industry: A flexible and knowledge based approach. In: New Advances in Information Systems and Technologies. Cham: Springer; 2016. pp. 455-464. DOI: 10.1007/978-3-319-31232-3_43

[53] Aizad T, Maganga O, Sumislawska M, Burnham KJ. Virtual enterprise process monitoring: An approach towards predictive industrial maintenance. Advances in Intelligent Systems and Computing. 2015;**1089**:285-293

[54] Romero D, Wuest T, Harik R, Thoben KD. Towards a cyber-physical PLM environment: The role of digital product models, intelligent products, digital twins, product avatars and digital shadows. IFAC. 2020;**53**(2):10911-10916. DOI: 10.1016/j.ifacol.2020.12.2829

[55] Pereira MM, Frazzon EM. A data-driven approach to adaptive synchronization of demand and supply in omni-channel retail supply chains. International Journal of Information Management. 2021;**57**:102165. DOI: 10.1016/j.ijinfomgt.2020.102165

[56] Giddaluru MP, Gao J. A data preparation and migration framework for implementing modular product structures in PLM. IFIP Advances in Information and Communication Technology. 2019;**565**:201-210

[57] Ali MM, Doumbouya MB, Louge T, Rai R, Karray MH. Ontology-based approach to extract product's design features from online customers' reviews. Computers in Industry. 2020;**116**:103175. DOI: 10.1016/j.compind.2019.103175

[58] Soegoto ES, Delvi J, Sunaryo A. Usage of E-commerce in Increasing Company Power and Sales. IOP Conference Series: Materials Science and Engineering. 2018;**407**(1)

[59] Brynjolfsson E, Hu YJ, Rahman MS. Competing in the age of omnichannel retailing. MIT Sloan Management Review. 2013;**54**(4):1-7

[60] Huovila H. Action Plan for Improving Omnichannel Customer Experience and Digital Lead Management. Tampere University of Technology. 2017

[61] Chen X, Wang X, Jiang X. The impact of power structure on the retail service supply chain with an O2O mixed channel. The Journal of the Operational Research Society. 2016;**67**(2):294-301. DOI: 10.1057/jors.2015.6

Section 2

Robust Control

Chapter 5

Hierarchical Sliding Mode Control for a 2D Ballbot That Is a Class of Second-Order Underactuated System

Dinh Ba Pham, Soon-Geul Lee, Thi Hang Bui and Tien Phat Truong

Abstract

2D Ballbot is an actual under-actuated system with second-order nonholonomic velocity constraints and input coupling case where only control input is employed to control two outputs of the system. Controlling such a system is not easy because it faces many changelings including nonlinearities, external disturbances, and uncertainties. This study proposed a robust control system for a Ballbot mobile robot. The proposed control scheme is constructed using the hierarchical sliding mode control technique. The kinematics and dynamics of the Ballbot are derived. A Lyapunov function is used to prove the stability of the closed-loop control system. The stabilizing and transferring problems are investigated through several simulations and experiments by using the actual Ballbot platform.

Keywords: sliding mode control, Ballbot, under-actuated system, Lyapunov analysis, stabilizing and transferring

1. Introduction

Ballbot has a body that balances on a single spherical wheel (ball). The robot uses a drive mechanism consisting of three omnidirectional wheels (OWs) for a ball [1] to ensure both stabilizing and transferring. The Ballbot can free travel in any direction on plat plane, even if the robot is crowded with people in a common human-coexisting environment.

An inherent property of the Ballbot system is a naturally nonlinear underactuated multi-input multi-output system, in which the number of the control input signal is less than the number of output signals. This property introduces the challenge related to Ballbot control design. Several control designs were accomplished using a simplified model (linear model) to overcome the complexities of the mathematical equations. The linear model is the first way to deal with the complexity of the mathematical equation of the Ballbot system. Reference [2] designed a control system based on a

IntechOpen

state feedback controller to control stabilizing and transferring. The experimental results showed the ability to stabilize and travel with loads in any direction. However, it can be led only to small external disturbances. Studies in [3, 4] employed a double-loop linear control system to stabilize a Ballbot. Their control system had two controllers: a PI inner-loop controller and a linear quadratic regulator (LQR) outer-loop controller. Kantor et al. [5] proposed a double-loop control system. Both inner-loop and outer-loop controllers were PID controllers. Sukvichai and Parnichkun [6] used a linear controller based on an LQR control scheme for a double-level Ballbot.

The second way to deal with the complexity of the mathematical equation of the Ballbot system is to use nonlinear and intelligent controls such as partial feedback linearization control (FLC), sliding mode control (SMC), or fuzzy control. Lotfiani et al. [7] employed an SMC controller along fuzzy trajectory planning to control a Ballbot to track the desired trajectory. Moreover, a collocated PFL method [8] was introduced for a Ballbot. The open-loop trajectory generation with the collocated PFL controller was simulated which showed a low position error.

Some researchers have studied intelligent control techniques. Reference [9] suggested an intelligent tracking control system combined with a dual Mamdani-type fuzzy control strategy and supervisory control technique for an omnidirectional spherical mobile platform. The experiment results showed the position tracking response of the robot. A fuzzy wavelet cerebellar-model-articulation controller [10] was proposed for a team of multiple Ballbots.

SMC is a well-known and robust nonlinear control scheme. To enhance robustness for both actuated [11] and under-actuated [12–19] systems, several controllers have been employed the SMC control method. The application of SMC for Ballbot control can be found in several previous studies. Ching-Wen et al. [20] introduced a hierarchical SMC based on backstepping to control stabilizing and agile trajectory tracking of a Ballbot with exogenous disturbance. Reference [21] enhanced Ching-Wen's backstepping SMC using interval type 2 fuzzy neural networks for motion control of a Ballbot with a four-motor inverse mouse ball-diving mechanism. However, these studies only presented numerical simulations.

As mentioned, the Ballbot is naturally an under-actuated system in which the number of the control inputs is less than the number of the outputs. The Ballbot system is nonlinear. Moreover, the control design problem also faces uncertainties, un-modeling, and disturbances. Thus, the SMC control scheme is a suitable approach for the Ballbot system.

In this study, we designed a robust controller to control a Ballbot. The proposed control scheme is a basis of the hierarchical SMC technique to execute two main tasks including stabilizing the body on the top of the ball and maneuvering the ball on the floor.

2. System model

2.1 Kinematic model

The general kinematics of the Ballbot is a relationship between $\dot{\boldsymbol{\Phi}}_k = \begin{bmatrix} \dot{\varphi}_x & \dot{\varphi}_y & \dot{\varphi}_z \end{bmatrix}^T$, the angular velocity of the ball relative to the body frame, and $\dot{\psi}_i$ $(i = 1, 2, 3)$, the angular velocities of the three OWs relative to the reference frame of the body. The kinematics relationship of dual-row OWs was discussed in [22]. For the Ballbot, the three OWs, a kind of single-row OW, are mounted on the body of the robot. The driving axes are separated by 120° in the x-y plane and each driving axis is sloped downward by α.

The main assumptions for simplification of the system are:

i. Body and floor are absolute solids.

ii. No slippage occurs on the surfaces of the ball-floor and the ball-OW.

iii. The points of contact of the ball-ground and the OWs-ball are points of contact.

The non-slip condition of the surface of the ball-OWs is met by requiring that the projection of velocities of the ball at all contact points in the actuation direction of the OW be the same and expressed mathematically as

$$\dot{\psi}_i r_w = \left(\dot{\boldsymbol{\phi}}_k \times \mathbf{P}_i\right)\mathbf{u}_{wi}, i = 1, 2, 3, \tag{1}$$

where $\mathbf{P}_i$ is the position vector from the center of the ball to the contact point i between the ball and the ith OW in the body frame as indicated in **Figure 1**, $\mathbf{u}_{wi}$ is the vector of direction cosines of the ith OW contact point velocities in the actuation directions, and r_w is the radius of the OW.

Therefore, the angular velocities of the OWs are expressed as

$$\dot{\psi}_i = \frac{1}{r_w}\left(\dot{\boldsymbol{\phi}}_k \times \mathbf{P}_i\right)\mathbf{u}_{wi}, i = 1, 2, 3. \tag{2}$$

As shown in **Figure 1**, the position vector of the contact point $\mathbf{P}_i$ can be defined by

$$\mathbf{P}_1 = \begin{bmatrix} r_k \sin\alpha & 0 & r_k \cos\alpha \end{bmatrix}^T, \tag{3a}$$

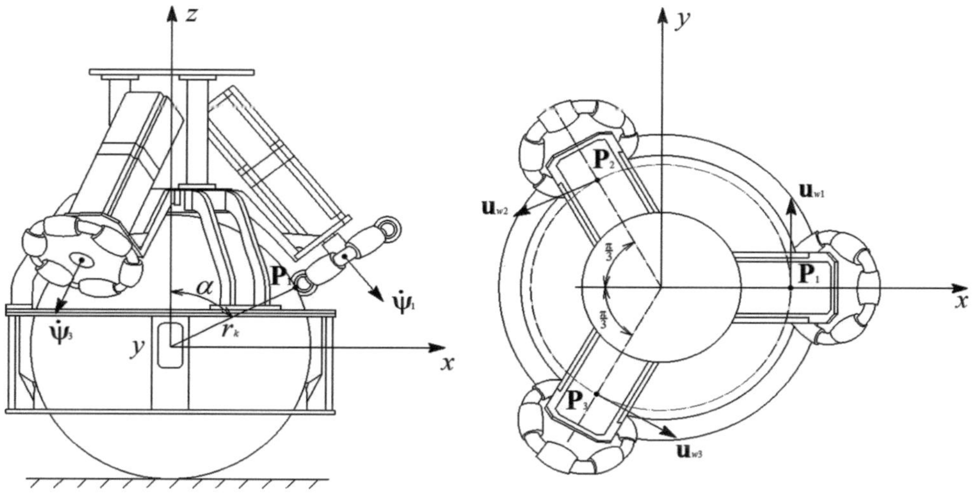

Figure 1.
Sketch of decomposed angular velocities of three OWs and a ball.

$$\mathbf{P}_{2,3} = \begin{bmatrix} -\frac{1}{2}r_k \sin\alpha & \pm\frac{\sqrt{3}}{2}r_k \sin\alpha & r_k \cos\alpha \end{bmatrix}^T. \tag{3b}$$

Furthermore, the direction cosine vectors that provide the corresponding direction of the speed at the contact point between the OWs and the ball are defined by

$$\mathbf{u}_{w1} = \begin{bmatrix} 0 & 1 & 0 \end{bmatrix}^T, \tag{4a}$$

$$\mathbf{u}_{w2,3} = \begin{bmatrix} \mp\frac{\sqrt{3}}{2} & -\frac{1}{2} & 0 \end{bmatrix}^T. \tag{4b}$$

Substituting (3a), (3b), (4a) and (4b) into (2) yields

$$\begin{bmatrix} \dot{\varphi}_x & \dot{\varphi}_y & \dot{\varphi}_z \end{bmatrix}^T = \mathbf{J}(\alpha)\begin{bmatrix} \dot{\psi}_1 & \dot{\psi}_2 & \dot{\psi}_3 \end{bmatrix}^T, \tag{5}$$

where $\mathbf{J}(\alpha)$ is the Jacobian matrix of the system

$$\mathbf{J}(\alpha) = \begin{bmatrix} -\frac{2r_w}{3r_k \cos\alpha} & \frac{r_w}{3r_k \cos\alpha} & \frac{r_w}{3r_k \cos\alpha} \\ 0 & -\frac{\sqrt{3}r_w}{3r_k \cos\alpha} & \frac{\sqrt{3}r_w}{3r_k \cos\alpha} \\ \frac{r_w}{3r_k \sin\alpha} & \frac{r_w}{3r_k \sin\alpha} & \frac{r_w}{3r_k \sin\alpha} \end{bmatrix}. \tag{6}$$

The Jacobian matrix (6) represents the effect of the zenith angle on the motion relationship between the ball and three single-row OWs.

2.2 Dynamic model

Figure 2 presents the planar model of the Ballbot. The system of the Ballbot is modeled as a rigid cuboid on top of a ball to get a simple calculation of the system parameters. The ball mass m_k and OW mass m_w are considered as point masses concentrated at their centers. m_a represents the equivalent mass of all components including the body, drive mechanism, and other parts. Chosen generalized coordinates of the Ballbot system include $x_k(t)$, $y_k(t)$, $\theta_x(t)$, and $\theta_y(t)$, here, $x_k(t)$, $y_k(t)$ denote

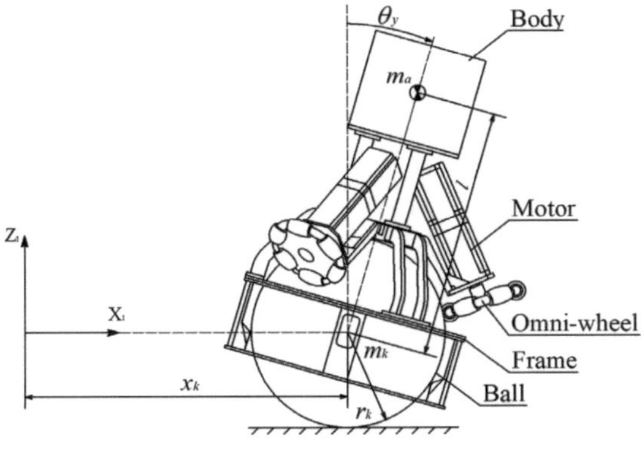

Figure 2.
Planar model of the main driving system of the Ballbot.

the position of the, $\theta_x(t)$, and $\theta_y(t)$ are the orientation of the body around the x- and y-axes. The control input is the resultant moment of the three actuators, whose x and y terms are τ_x and τ_y.

The Ballbot system dynamics are described by obtaining the kinetic and potential of the ball, body, and OWs. For the ball,

$$T_{kx} = \frac{1}{2}\left(m_k + \frac{I_k}{r_k^2}\right)\dot{y}_k^2, V_{kx} = 0, \tag{7}$$

where r_k is the ball radius and I_k is the ball momentum inertia.

The OWs are attached to the body of the Ballbot. Thus, only the rotational motion of the OWs has to be calculated, that is,

$$T_{wx} = \frac{3I_w \cos^2\alpha}{4r_w^2}\left(\dot{y}_k + r_k\dot{\theta}_x\right)^2, V_{wx} = 0, \tag{8}$$

where r_w is the radius of each OW, α represents the zenith angle, and I_w indicates the momentum inertia of each OW.

For the body,

$$T_{ax} = \frac{1}{2}I_x\dot{\theta}_x^2 + \frac{1}{2}m_a\left(\dot{y}_k - l\dot{\theta}_x\cos\theta_x\right)^2 + \frac{1}{2}m_a l^2\dot{\theta}_x^2\sin^2\theta_x, V_{ax} = m_a g l\cos\theta_x, \tag{9}$$

where I_x is the body momentum inertia around the x-axis, l represents the distance from the ball center to the body mass center, and g denotes the gravitational acceleration.

In the y-z plane, the generalized coordinates of the Ballbot system are defined as $\mathbf{q}_x = \begin{bmatrix} y_k & \theta_x \end{bmatrix}^T$ and the friction vector

$$\mathbf{D}(\dot{\mathbf{q}}_x) = \begin{bmatrix} b_y\dot{y}_k & b_{rx}\dot{\theta}_x \end{bmatrix}^T, \tag{10}$$

where b_y and b_{rx} are the viscous damping coefficients that model the ball-floor friction and ball-body friction in the y-z plane, respectively.

The Lagrangian function L_x is expressed as

$$\begin{aligned} L_x &= T_x - V_x = T_{kx} + T_{wx} + T_{ax} - (V_{kx} + V_{wx} + V_{ax}) \\ &= \begin{pmatrix} \frac{1}{2}\left(m_k + \frac{I_k}{r_k^2}\right)\dot{y}_k^2 + \frac{3I_w\cos^2\alpha}{4r_w^2}\left(\dot{y}_k + r_k\dot{\theta}_x\right)^2 \\ +\frac{1}{2}I_x\dot{\theta}_x^2 + \frac{1}{2}m_a\left(\dot{y}_k - l\dot{\theta}_x\cos\theta_x\right)^2 + \frac{1}{2}m_a l^2\dot{\theta}_x^2\sin^2\theta_x - m_a g l\cos\theta_x \end{pmatrix}, \end{aligned} \tag{11}$$

The Euler-Lagrange equation that describes the motion of the Ballbot is

$$\frac{d}{dt}\left(\frac{\partial L_x}{\partial \dot{\mathbf{q}}_x}\right) - \frac{\partial L_x}{\partial \mathbf{q}_x} = \frac{1}{r_w}\begin{bmatrix} 1 \\ r_k \end{bmatrix}\tau_x - \mathbf{D}(\dot{\mathbf{q}}_x). \tag{12}$$

From (12), the dynamic equations on the y-z plane are expressed as follows:

$$\ddot{y}_k a_1 + (a_4 - a_3\cos\theta_x)\ddot{\theta}_x + a_3\dot{\theta}_x^2\sin\theta_x + b_y\dot{y}_k = r_w^{-1}\tau_x, \tag{13}$$

$$(a_4 - a_3\cos\theta_x)\ddot{y}_k + \ddot{\theta}_x a_2 + b_{rx}\dot{\theta}_x - a_5\sin\theta_x = r_k r_w^{-1}\tau_x, \tag{14}$$

where $a_1 = m_k + \frac{I_{kx}}{r_k^2} + m_a + \frac{3I_w\cos^2\alpha}{2r_w^2}$, $a_2 = m_a l^2 + \frac{3I_w r_k^2\cos^2\alpha}{2r_w^2} + I_x$, $a_3 = m_a l$, $a_4 = \frac{3I_w\cos^2\alpha}{2r_w^2} r_k$, $a_5 = m_a g l$, I_k refers to the ball momentum inertia, I_w is the OW momentum inertia, I_x denotes the body momentum inertia about the x-axis, b_y and b_{rx} are the viscous damping coefficients.

Then, (13) and (14) can be rewritten as follows:

$$\ddot{y}_k = F_{x1}(\mathbf{q}_x, \dot{\mathbf{q}}_x) + G_{x1}(\mathbf{q}_x)\tau_x, \tag{15}$$

$$\ddot{\theta}_x = F_{x2}(\mathbf{q}_x, \dot{\mathbf{q}}_x) + G_{x2}(\mathbf{q}_x)\tau_x, \tag{16}$$

where $F_{x1}(\mathbf{q}_x, \dot{\mathbf{q}}_x) = A_x^{-1}\left[(a_3\cos\theta_x - a_4)(a_5\sin\theta_x - b_{rx}\dot{\theta}_x) - a_2\left(a_3\dot{\theta}_x^2\sin\theta_x + b_y\dot{y}_k\right)\right]$,

$$G_{x1}(\mathbf{q}_x) = A_x^{-1} r_{\mathrm{w}}^{-1}(a_2 + a_3 r_k\cos\theta_x - a_4 r_k),$$

$$F_{x2}(\mathbf{q}_x, \dot{\mathbf{q}}_x) = A_x^{-1}\left[(a_4 - a_3\cos\theta_x)\left(a_3\dot{\theta}_x^2\sin\theta_x + b_y\dot{y}_k\right) + a_1(a_5\sin\theta_x - b_{rx}\dot{\theta}_x)\right],$$

$$G_{x2}(\mathbf{q}_x) = A_x^{-1} r_w^{-1}(a_3\cos\theta_x - a_4 + a_1 r_k),$$

$$A_x = a_1 a_2 - (a_4 - a_3\cos\theta_x)^2.$$

In the x-z plane, the mathematical equations describe the Ballbot system dynamics as follows:

$$\ddot{x}_x b_1 + (b_4\cos\theta_y - b_3)\ddot{\theta}_y - b_4\dot{\theta}_y^2\sin\theta_y + b_x\dot{x}_k = -r_w^{-1}\tau_y, \tag{17}$$

$$(b_4\cos\theta_y - b_3)\ddot{x}_k + \ddot{\theta}_y b_2 - b_5\sin\theta_y + b_{ry}\dot{\theta}_y = r_k r_w^{-1}\tau_y, \tag{18}$$

where $b_1 = m_k + \frac{I_k}{r_k^2} + \frac{3I_w\cos^2\alpha}{2r_w^2} + m_a$, $b_2 = \frac{3I_w r_k^2\cos^2\alpha}{2r_w^2} + m_a l^2 + I_y$, $b_3 = \frac{3I_w\cos^2\alpha}{2r_w^2} r_k$, $b_4 = m_a l$, $b_5 = m_a g l$, I_y represents the body momentum inertia about the y-axis, b_x and b_{ry} are the viscous damping coefficients.

In the x-z plane, minimal coordinates of the Ballbot system denotes $\mathbf{q}_y = \begin{bmatrix} x_k & \theta_y \end{bmatrix}^T$, then (17) and (18), are rewritten as follows:

$$\ddot{x}_k = F_{y1}\left(\mathbf{q}_y, \dot{\mathbf{q}}_y\right) + G_{y1}\left(\mathbf{q}_y\right)\tau_y, \tag{19}$$

$$\ddot{\theta}_y = F_{y2}\left(\mathbf{q}_y, \dot{\mathbf{q}}_y\right) + G_{y2}\left(\mathbf{q}_y\right)\tau_y, \tag{20}$$

where $F_{y1}\left(\mathbf{q}_y, \dot{\mathbf{q}}_y\right) = A_y^{-1}\left[b_2\left(b_4\sin\theta_y\dot{\theta}_y^2 - b_x\dot{x}_k\right) + (b_3 - b_4\cos\theta_y)(b_5\sin\theta_y - b_{ry}\dot{\theta}_y)\right]$,

$$G_{y1}\left(\mathbf{q}_y\right) = -A_y^{-1} r_w^{-1}(b_2 - b_3 r_k + b_4 r_k\cos\theta_y),$$

$$F_{y2}\left(q_y, \dot{q}_y\right) = A_y^{-1}\left[(b_3 - b_4\,cos\,\theta_y)\left(b_4\,sin\,\theta_y\dot{\theta}_y^2 - b_x\dot{x}_k\right) + b_1(b_5\,sin\,\theta_y - b_{ry}\dot{\theta}_y)\right],$$

$$G_{y2}\left(\mathbf{q}_y\right) = A_y^{-1} r_w^{-1}\left(r_k b_1 - b_3 + b_4 \cos\theta_y\right),$$
$$A_y = b_1 a_2 - \left(b_4 \cos\theta_y - b_3\right)^2.$$

3. Control system design

3.1 Hierarchical sliding mode control design

It can be assumed that all state variables are measurable. To design this controller in the y-z plane, first we introduce a suitable pair of SMSs

$$\begin{cases} s_{x1} = c_{x1} e_{x1} + \dot{e}_{x1} \\ s_{x2} = c_{x2} e_{x2} + \dot{e}_{x2} \end{cases}, \tag{21}$$

where c_{x1} and c_{x2} are positive constants, e_{x1} and e_{x2} are tracking errors

$$\begin{cases} e_{x1} = y_k - y_{kd} \\ e_{x2} = \theta_x - \theta_{xd} \end{cases}, \tag{22}$$

where $y_{kd} = const$ denotes the desired position of the ball and θ_{xd} is the desired tilt angle about the x-axis of the body.

When the Ballbot balances, it means that the desired tilt angle $\theta_{xd} = 0$. Then, (21) can be rewritten as

$$\begin{cases} s_{x1} = c_{x1}\left(y_k - y_{kd}\right) + \dot{y}_k \\ s_{x2} = c_{x2}\theta_x + \dot{\theta}_x \end{cases}. \tag{23}$$

Let $\dot{s}_{x1} = 0$ and $\dot{s}_{x2} = 0$, the equivalent control laws of the two subsystems can be gotten as

$$\tau_{xeq1} = -G_{x1}^{-1}\left(\mathbf{q}_x\right)\left[c_{x1}\dot{y}_k + F_{x1}\left(\mathbf{q}_x, \dot{\mathbf{q}}_x\right)\right], \tag{24}$$
$$\tau_{xeq2} = -G_{x2}^{-1}\left(\mathbf{q}_x\right)\left[c_{x2}\dot{\theta}_x + F_{x2}\left(\mathbf{q}_x, \dot{\mathbf{q}}_x\right)\right]. \tag{25}$$

The hierarchical SMC law is deduced as follows. The first layer SMS is defined as $S_{x1} = s_{x1}$. For the first layer SMS, the SMC law and the Lyapunov function are defined as

$$\tau_{x1} = \tau_{xeq1} + \tau_{xsw1}, \tag{26}$$

and

$$V_{x1}(t) = 0.5 S_{x1}^2, \tag{27}$$

where τ_{xsw1} is the switch control part of the first layer SMC. Differentiate $V_{x1}(t)$ with respect to time t

$$\dot{V}_{x1}(t) = S_{x1}\dot{S}_{x1}. \tag{28}$$

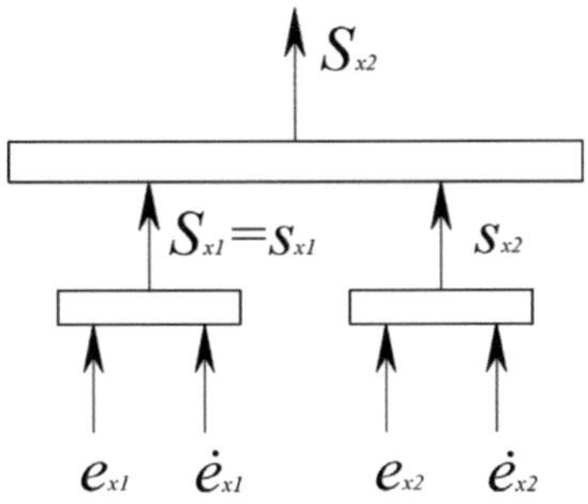

Figure 3.
Structure of hierarchical sliding mode surfaces.

Let

$$\dot{S}_{x1} = -k_{x1}S_{x1} - \eta_{x1}sign(S_{x1}), \tag{29}$$

where k_{x1} and η_{x1} are positive constants.
The first layer SMC law can be deduced from Eqs. (26) and (27), that is,

$$\tau_{x1} = \tau_{xeq1} + G_{x1}^{-1}(\mathbf{q}_x)\dot{S}_{x1}. \tag{30}$$

The second layer SMS is constructed based on the first layer SMS S_1 and s_2, as shown in **Figure 3**.

$$S_{x2} = \alpha_x S_{x1} + s_{x2}, \tag{31}$$

where α_x is the sliding mode parameter.
For the second layer SMS, the SMC law and the Lyapunov function are defined as

$$\tau_{x2} = \tau_{x1} + \tau_{xeq2} + \tau_{xsw2}, \tag{32}$$

and

$$V_{x2}(t) = 0.5S_{x2}^2, \tag{33}$$

where τ_{xsw2} is the switch control part of the second layer SMC.
Differentiating $V_{x2}(t)$ with respect to time t yields

$$\dot{V}_{x2}(t) = S_{x2}\dot{S}_{x2}. \tag{34}$$

Let

$$\dot{S}_{x2} = -k_{x2}S_{x2} - \eta_{x2}sign(S_{x2}), \tag{35}$$

where k_{x2} and η_{x2} are positive constants.
The total control law of the presented hierarchical SMC can be deduced as follows:

$$\tau_{x2} = \frac{\alpha_x G_{x1}(\mathbf{q}_x)\tau_{xeq1} + G_{x2}(\mathbf{q}_x)\tau_{xeq2} + \dot{S}_{x2}}{\alpha_x G_{x1}(\mathbf{q}_x) + G_{x2}(\mathbf{q}_x)}. \tag{36}$$

Similarly, the total control law of the hierarchical SMC in the x-z plane also given as

$$\tau_{y2} = \frac{\alpha_y G_{y1}\left(\mathbf{q}_y\right)\tau_{yeq1} + G_{y2}\left(\mathbf{q}_y\right)\tau_{yeq2} + \dot{S}_{y2}}{\alpha_y G_{y1}\left(\mathbf{q}_y\right) + G_{y2}\left(\mathbf{q}_y\right)}. \tag{37}$$

3.2 Stability analysis

Theorem 1: If considering the total control law (36) and the SMSs (23) and (31) for the system dynamics (15) and (16), then SMSs, S_{x1} and S_{x2} are asymptotically stable.

Proof: Integrating both sides (34) with respect to time obtains

$$\int_0^t \dot{V}_{x2} d\tau = \int_0^t (-\eta_{x2}|S_{x2}| - k_{x2}S_{x2}^2) d\tau, \tag{38}$$

Then

$$V_{x2}(t) - V_{x2}(0) = \int_0^t (-\eta_{x2}|S_{x2}| - k_{x2}S_{x2}^2) d\tau, \tag{39}$$

It can be found that

$$V_{x2}(0) = V_{x2}(t) + \int_0^t (\eta_{x2}|S_{x2}| + k_{x2}S_{x2}^2) d\tau \geq \int_0^t (\eta_{x2}|S_{x2}| + k_{x2}S_{x2}^2) d\tau. \tag{40}$$

Therefore, it can be achieved that

$$\lim_{t\to\infty} \int_0^t (\eta_{x2}|S_{x2}| + k_{x2}S_{x2}^2) d\tau \leq V_{x2}(0) < \infty.$$

By using Barbalat's lemma [23], we can obtain that if $t \to \infty$ then $\eta_{x2}|S_{x2}| + k_{x2}S_{x2}^2 \to 0$. Then, $\lim_{t\to\infty} S_{x2} = 0$.

By applying Barbalat's lemma, we can get $\lim_{t\to\infty} S_{x2} = 0$.

Thus, both S_{x1} and S_{x2} are asymptotically stable.

Theorem 2: If considering the control law (36) and the SMSs of (23) for the system dynamics (15) and (16), then SMSs, s_{x1} and s_{x2}, are also asymptotically stable.

Proof: From Theorem 1, the SMS of the ball subsystem dynamics is asymptotically stable.

Now, we will prove that the SMS of the body subsystem dynamics is asymptotically stable. Limiting of both sides of (31) obtains

$$\lim_{t\to\infty} S_{x2} = \lim_{t\to\infty} (\alpha_x S_{x1} + s_{x2}) = \alpha_x \left(\lim_{t\to\infty} S_{x1}\right) + \lim_{t\to\infty} s_{x2} = \lim_{t\to\infty} s_{x2}. \tag{41}$$

The result of (41) shows $\lim_{t\to\infty} s_{x2} = \lim_{t\to\infty} S_{x2} = 0$. It demonstrates that the SMS of the body subsystem dynamics is asymptotically stable. Thus, the all SMSs of the subsystems are asymptotically stable.

Theorem 3: If considering total control law (36) and the SMSs (23) for the system dynamics (15) and (16), assuming that all the state variables are equivalent infinitesimal, then the parameter boundary of the SMS is $0 < c_{x1} < \left| \lim_{\mathbf{x} \to \mathbf{0}} \left(\frac{F_{x1}(\mathbf{q}_x, \dot{\mathbf{q}}_x)}{\dot{y}_k} \right) \right|$ and $0 < c_{x2} < \left| \lim_{\mathbf{x} \to \mathbf{0}} \left(\frac{F_{x2}(\mathbf{q}_x, \dot{\mathbf{q}}_x)}{\dot{\theta}_x} \right) \right|$, where $\mathbf{x} = \begin{bmatrix} \mathbf{q}_x & \dot{\mathbf{q}}_x \end{bmatrix}$.

Proof: By solving $s_{xi} = 0$, the lower boundary of c_{xi} can be obtained

$$\begin{cases} s_{x1} = c_{x1}(y_k - y_{kd}) + \dot{y}_k = 0 \\ s_{x2} = c_{x2}\theta_x + \dot{\theta}_x = 0 \end{cases}. \tag{42}$$

The eigenvalue of (42) should be negative for meeting the system stability. Thus, the lower boundary of c_{xi} is $c_{xi} > 0$.

The upper boundary of c_{xi} can be gotten from $\dot{s}_{xi} = 0$

$$\begin{cases} \dot{s}_{x1} = c_{x1}\dot{y}_k + F_{x1}(\mathbf{q}_x, \dot{\mathbf{q}}_x) + G_{x1}(\mathbf{q}_x)\tau_{xeq1} = 0 \\ \dot{s}_{x2} = c_{x2}\dot{\theta}_x + F_{x2}(\mathbf{q}_x, \dot{\mathbf{q}}_x) + G_{x2}(\mathbf{q}_x)\tau_{xeq2} = 0 \end{cases}. \tag{43}$$

Therefore,

$$\begin{cases} c_{x1} = \left| \dfrac{(F_{x1}(\mathbf{q}_x, \dot{\mathbf{q}}_x) + G_{x1}(\mathbf{q}_x)\tau_{xeq1})}{\dot{y}_k} \right| \\ c_{x2} = \left| \dfrac{(F_{x2}(\mathbf{q}_x, \dot{\mathbf{q}}_x) + G_{x2}(\mathbf{q}_x)\tau_{xeq2})}{\dot{\theta}_x} \right| \end{cases}. \tag{44}$$

Further, it can be got

$$\begin{cases} c_{x1} < \dfrac{(|F_{x1}(\mathbf{q}_x, \dot{\mathbf{q}}_x)| + |G_{x1}(\mathbf{q}_x)\tau_{xeq1}|)}{|\dot{y}_k|} \\ c_{x2} < \dfrac{(|F_{x2}(\mathbf{q}_x, \dot{\mathbf{q}}_x)| + |G_{x2}(\mathbf{q}_x)\tau_{xeq2}|)}{|\dot{\theta}_x|} \end{cases}. \tag{45}$$

When the state of the subsystem keeps sliding on its SMS and converges to the neighborhood of the control objective, the system can be treated as an autonomous one. Thus, $\tau_{xeqi} = 0$ and the following inequation is gotten $0 < c_{x1} < \left| \lim_{\mathbf{x} \to \mathbf{0}} \left(\frac{F_{x1}(\mathbf{q}_x, \dot{\mathbf{q}}_x)}{\dot{y}_k} \right) \right|$ and $0 < c_{x2} < \left| \lim_{\mathbf{x} \to \mathbf{0}} \left(\frac{F_{x2}(\mathbf{q}_x, \dot{\mathbf{q}}_x)}{\dot{\theta}_x} \right) \right|$.

The block diagram of the control system is shown in **Figure 4**. With the real motor torques τ_1, τ_2, and τ_3 as functions of the equivalent torques about the x- and y-axes, τ_x and τ_y, in matrix form, yield [24].

$$\begin{bmatrix} \tau_1 \\ \tau_2 \\ \tau_3 \end{bmatrix} = \begin{bmatrix} \dfrac{2}{3\cos\alpha} & -\dfrac{1}{3\cos\alpha} & -\dfrac{1}{3\cos\alpha} \\ 0 & \dfrac{\sqrt{3}}{3\cos\alpha} & -\dfrac{\sqrt{3}}{3\cos\alpha} \end{bmatrix}^T \begin{bmatrix} \tau_x \\ \tau_y \end{bmatrix}. \tag{46}$$

The corresponding simulation and experiment results will be given in the subsequent sections.

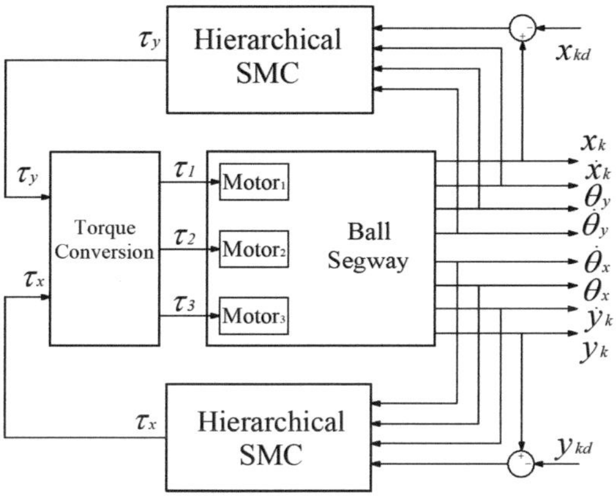

Figure 4.
Block diagram of the control system.

4. Simulation result

A numerical simulation investigates the stabilize, robustness, and transfer of the proposed controller for the Ballbot. The decoupled dynamics (15) and (16), and (19) and (20) with the proposed control schemes (36) and (37) are modeled in Matlab/Simulink real-time environment with ODE45 and a sampling time of 0.01 seconds.

The parameters of the Ballbot for both simulation and experiment are shown in **Table 1**. For the numerical simulation, the control parameters are tuned by the trial-and-error method and then selected as in **Table 1**.

Various simulations are conducted by considering the stabilizing controls with an initial nonzero tilt angle and with external disturbances and tracking control.

4.1 Stabilizing control with nonzero initial tilt angles

In this simulation, the tilt angles about x- and y-axes are initialized as 6.3° and −6.5°, respectively for checking the behavior. Simulation results are shown in **Figure 5**. The tilt angles responses and the angular velocities are depicted in **Figure 5(a)** and **(b)**, respectively. The position of the ball is shown in **Figure 5(c)**. **Figure 5(d)** shows the curves of the control inputs.

These numerical results demonstrate that the proposed robust controller enables to maintain stabilizing of the Ballbot.

System parameters	Control gains
$m_a = 116\text{kg}$, $I_x = 16.25\text{kgm}^2$, $I_y = 15.85\text{kgm}^2$, $r_w = 0.1\text{m}$, $l = 0.23\text{m}$, $I_w = 0.26\text{kgm}^2$, $r_k = 0.19\text{m}$, $m_k = 11.4\text{kg}$, $I_k = 0.165\text{kgm}^2$, $b_x = b_y = 5\text{Ns/m}$, $b_{rx} = b_{ry} = 3.68\text{Nms/rad}$, $\alpha = 56°$.	$c_{x1} = 0.01$, $c_{x2} = 35$, $\alpha_x = 0.05$, $\eta_{x2} = 0.1$, $k_{x2} = 10$; $c_{y_1} = 0.01$, $c_{y_2} = 17$, $\alpha_y = 0.05$, $\eta_{y2} = 0.1$, $k_{y2} = 10$

Table 1.
System parameters and control gains.

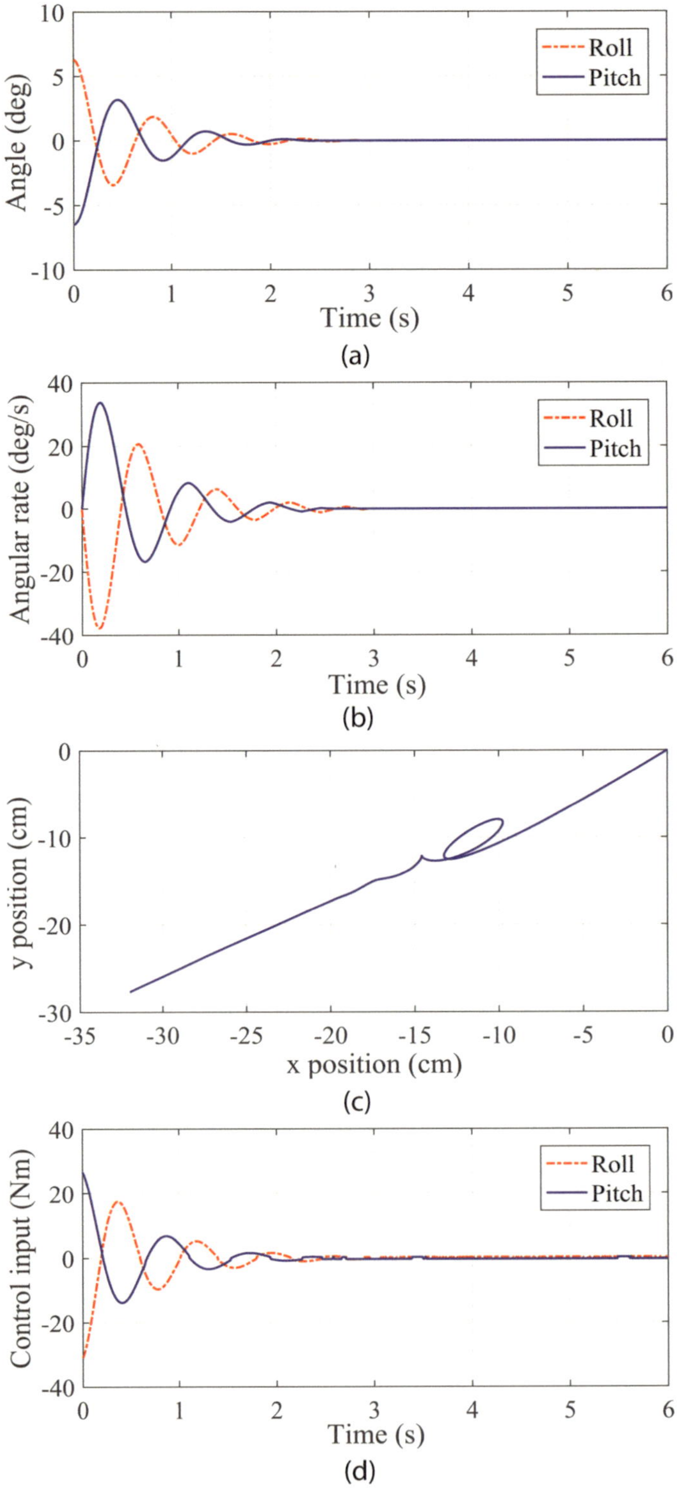

Figure 5.
Simulation results of stabilizing while station-keeping. (a) Tilt angles of the body. (b) Angular velocities of the body. (c) Trajectory of the contact point between the ball and the floor. (d) Control inputs.

4.2 Stabilizing control with external disturbances

In the second simulation, the stabilizing control of the Ballbot is investigated under an external disturbance. An external force of 300 N is applied to the Ballbot system at the sixth second while the Ballbot is stabilizing on the floor. As the results, the body tilt angles and angular tilt angles are depicted in **Figure 6(a)** and **(b)**, respectively. Under the external disturbance, The Ballbot cannot maintain its original position of $(x_k, y_k) = (0, 0)$. Instead, the robot has traveled to a new position of $(x_k, y_k) = (-7.4\text{cm}, 3.1\text{cm})$ as shown in **Figure 6(c)**. The control inputs by the proposed control are converged to zeros after 2 seconds, as depicted in **Figure 6(d)**.

4.3 Tracking control

In the third simulation, the Ballbot is commanded to track a rectangular trajectory with a dimension of 75 cm $\times$ 90 cm in 40 s. The system responses of the tilt angles of the body are shown in **Figure 7(b)**. As indicated in **Figure 7(a)**, the proposed control system performs well in tracking desired rectangular trajectory.

5. Experiment result

In this section, several experiments on the actual Ballbot platform (**Figure 8**) are implemented to further verify the performance of the proposed control system. Especially, experiment for the robustness of the controller is executed under external disturbances.

The robot control algorithm is programmed with multithread tasks so that the control period is set to 15 ms. The program also consists of the torque conversion and kinematic model [1] to estimate the position and velocity of the ball.

An inertial measurement unit (IMU) is utilized to measure the orientation and angular rates of the Ballbot. The IMU includes an accelerometer and a gyro sensor. Three encoders with a resolution of 4000 counts/rev are also utilized to obtain the position of the ball. Full state variables of the Ballbot system can be obtained based on the kinematics and the sensor fusion.

The drive mechanism is equipped with brushless DC actuators with a continuous torque of 0.28 Nm and gearboxes with a ratio of 1:4 for driving the ball. Two 48 V lithium battery packs supply power for the actuators and other devices with a working time of several hours.

5.1 Stabilizing control with an initial nonzero tilt angle

This experiment investigates the stabilizing performance of the proposed hierarchical SMC with an actual Ballbot on the flat floor.

The initial position of the Ballbot is set as the origin point and the robot is set at 6.3° in roll angle and $-$ 6.5° in pitch angle. The stabilizing responses of the control system are shown in **Figure 9**. The tilt angles of the body are presented in **Figure 9(a)** in which the steady-state is 1.5 seconds and the steady-state errors of the roll and pitch are 0.4° and 0.5°, respectively. While the angular rates of the body are shown in **Figure 9(b)**. The proposed control system successfully controls the movement of the ball from an origin

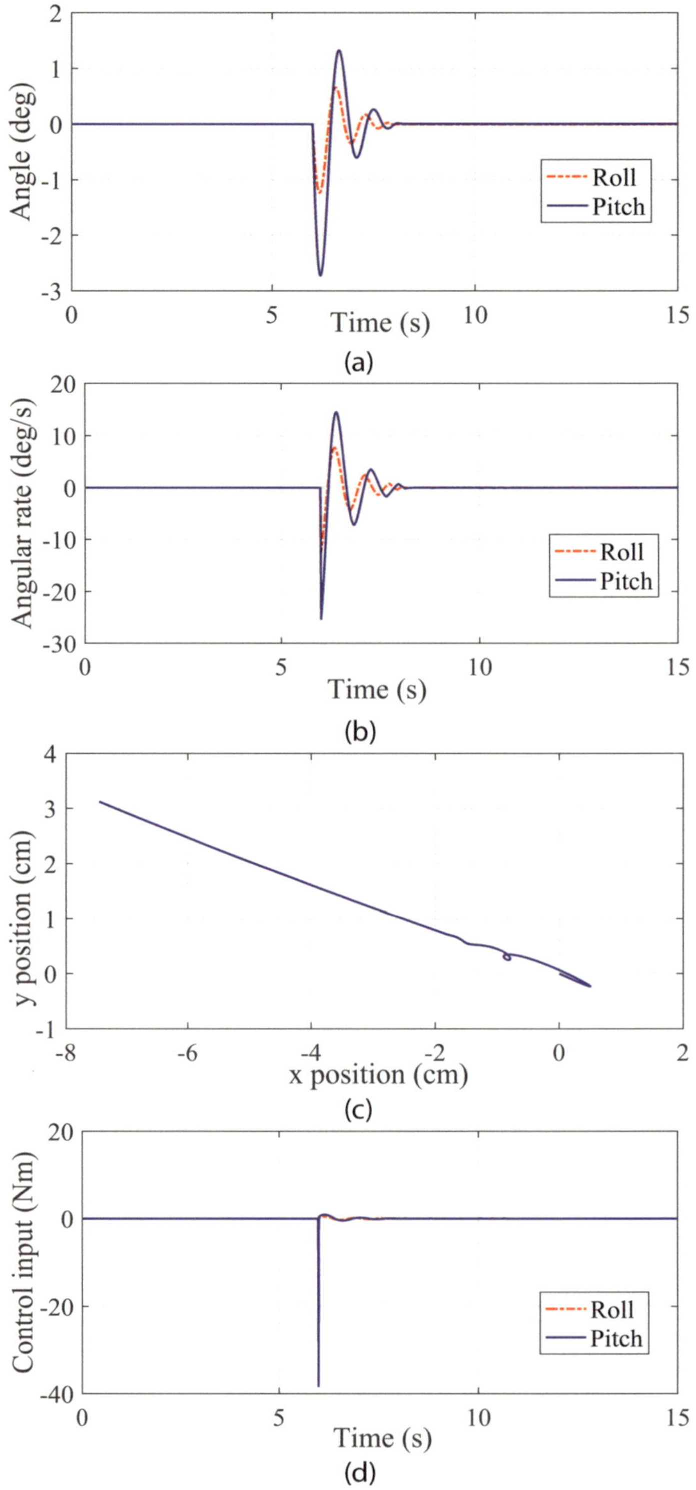

Figure 6.
Simulation results of robustness control. (a) Tilt angles of the body. (b) Angular velocities of the body. (c) Trajectory of the contact point between the ball and floor. (d) Control inputs.

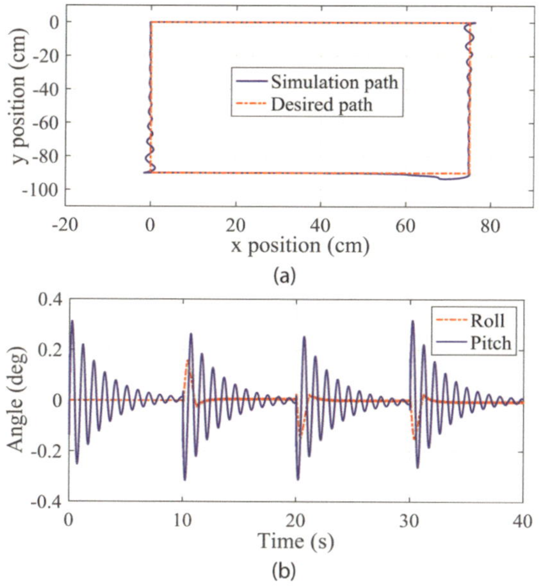

Figure 7.
Simulation results of tracking a rectangular path. (a) Trajectory of the contact point between the ball and floor. (b) Tilt angles of the body.

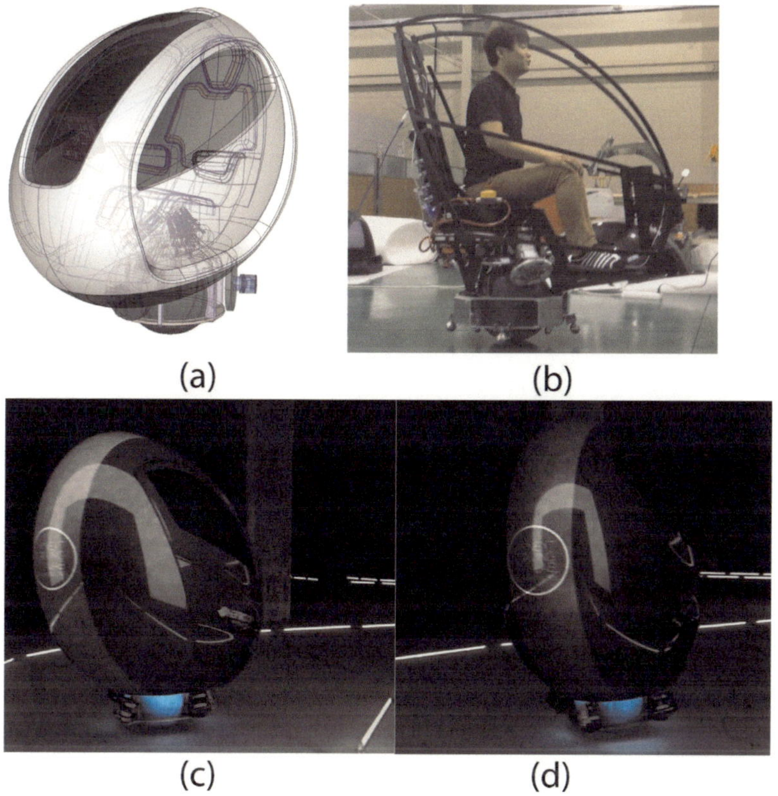

Figure 8.
The real Ballbot that is running. (a) Schematic design. (b) Without the cover. (c) Ballbot running. (d) Ballbot running.

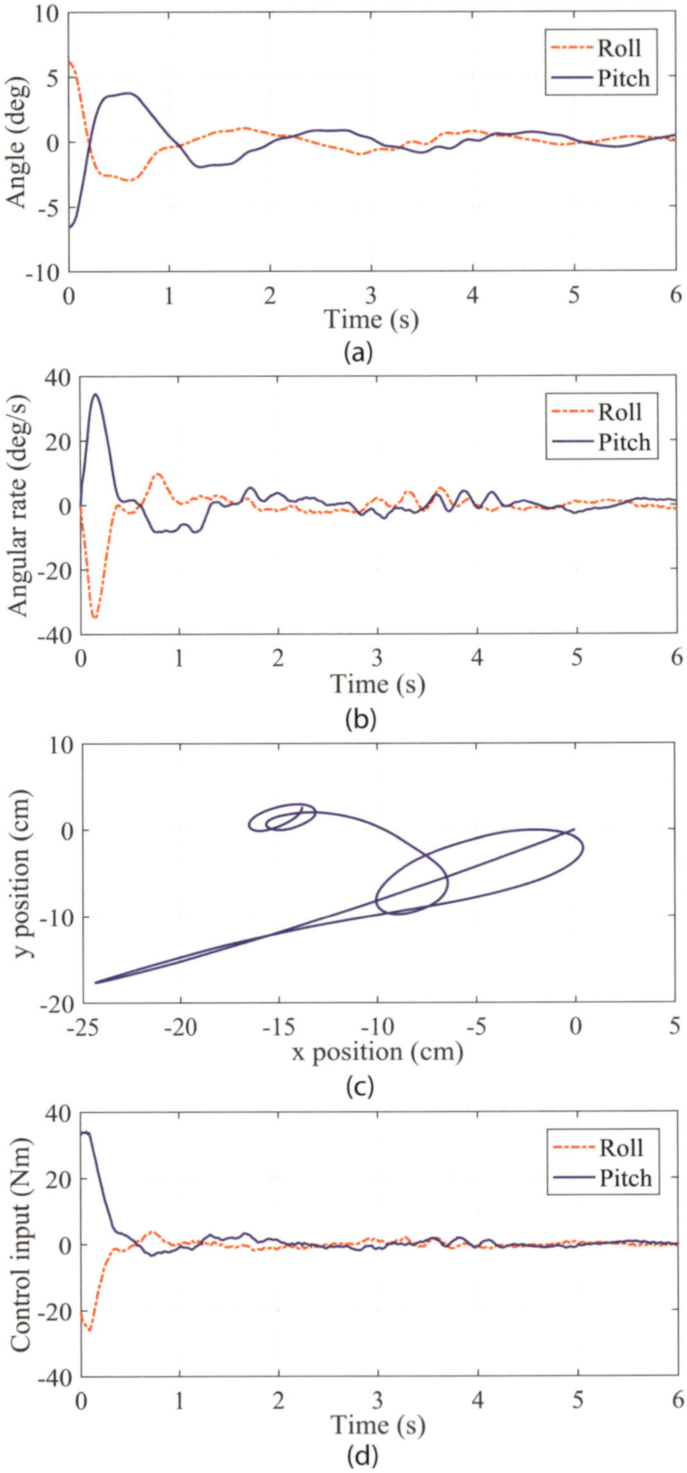

Figure 9.
Experiment results of stabilizing and station-keeping. (a) Tilt angles of the body. (b) Angular velocities of the body. (c) Trajectory of the contact point between the ball and floor. (d) Control inputs.

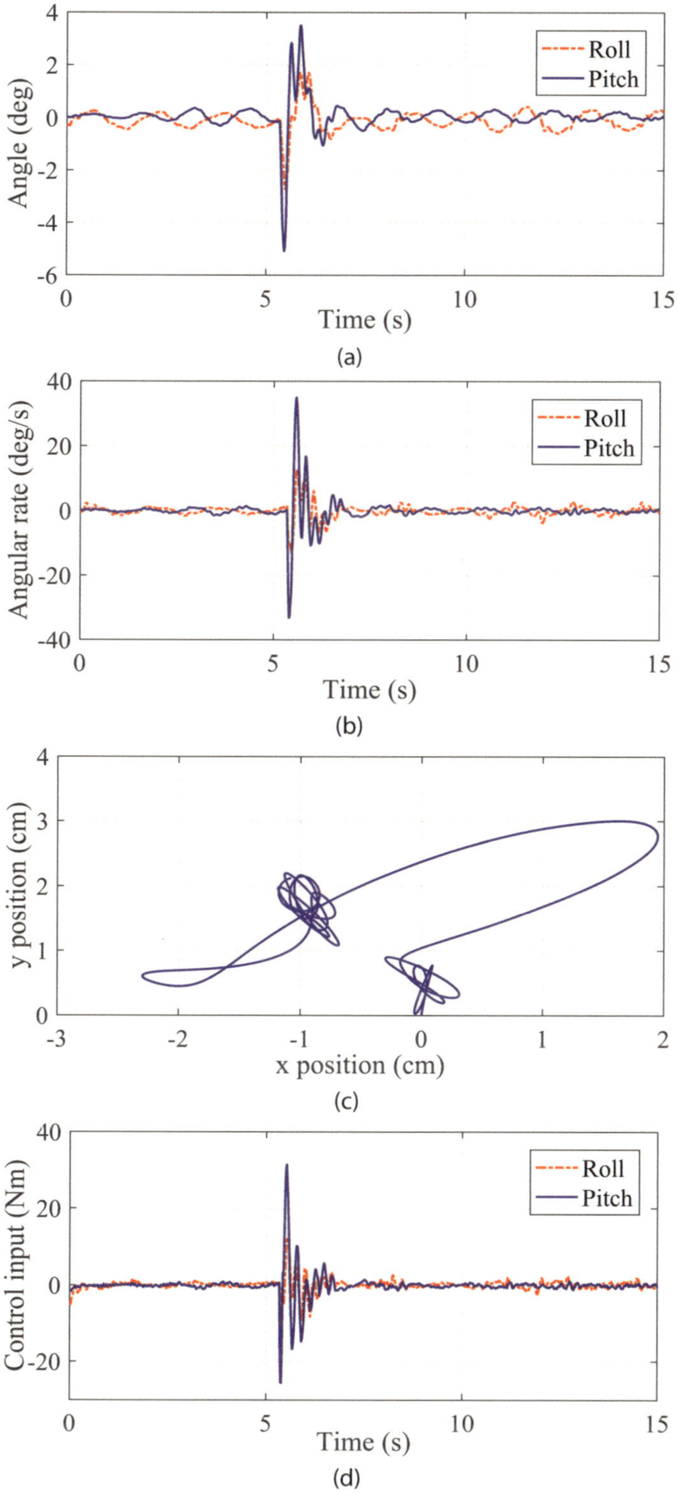

Figure 10.
Experiment results of robustness control performance. (a) Tilt angles of the body. (b) Angular velocities of the body. (c) Trajectory of the contact point between the ball and floor. (d) Control inputs.

point to the new point of $(x_k, y_k) = (-14 \text{ cm}, 2 \text{ cm})$ as shown in **Figure 9(c)**. The control input of the proposed control scheme is shown in **Figure 9(d)**.

5.2 Stabilizing control under an external disturbance

The robustness performance of the proposed hierarchical SMC is evaluated by applying the external disturbances to the robot. The experimental scenario is set as: at the beginning, the robot is stabilizing at the origin position. Then the Ballbot is kicked. The amount of the kick is about 300 N.

The tilt angles and angular rate of the body are shown in **Figure 10(a)** and **(b)**. **Figure 10(c)** shows the ball response along the x- and y-axes. Under the kick, the robot moves from the origin position of $(x_k, y_k) = (0, 0)$ to the new position of $(x_k, y_k) = (-0.92 \text{ cm}, 1.65 \text{ cm})$ and then stabilizes at the new position. Torque control input responses are also shown in **Figure 10(d)** to keep the stabilizing of the Ballbot.

5.3 Tracking control

In this experiment, the Ballbot is commanded to track the desired rectangular trajectory with a dimension of 75 cm × 90 cm within 40 seconds. The system response is presented in **Figure 11**. **Figure 11(a)** shows the trajectory of the ball on the floor.

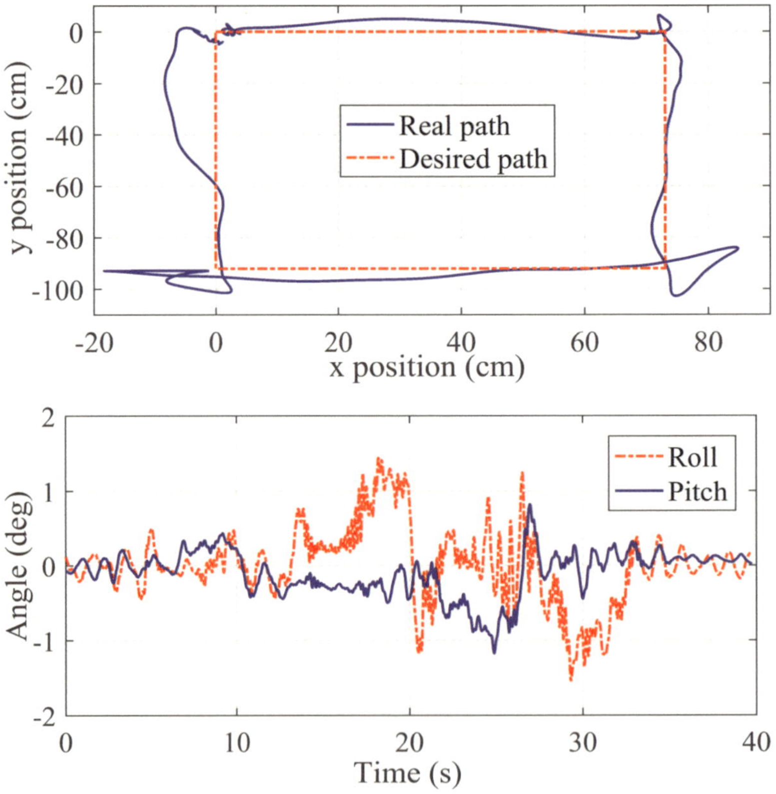

Figure 11.
Experimental results of tracking a rectangular path. (a) Trajectory of the contact point between the ball and floor. (b) Tilt angles of the body.

There is some error while the robot tries to track the rectangular desired trajectory. The position error occurs due to uncertainties, an un-modeling system.

The results demonstrate hierarchical SMC behaviors in stabilizing and transferring control of the Ballbot.

6. Conclusions

In this study, a 2D model of the Ballbot is obtained using the Euler-Lagrange equation and the decoupling method. A robust nonlinear controller based on the hierarchical SMC technique is designed for the Ballbot to control stabilizing and transferring. The capability of the closed-loop system with the hierarchical SMC is achieved using the Lyapunov function. The performance and robustness of the hierarchical SMC are examined under several tests in both simulation and experiment. The simulation and experimental results demonstrate the capabilities of the proposed controller for stabilizing and trajectory tracking.

Author details

Dinh Ba Pham[1]*, Soon-Geul Lee[2], Thi Hang Bui[3] and Tien Phat Truong[3]

1 Mechanical Engineering Department, Vietnam Maritime University, Haiphong, Vietnam

2 Mechanical Engineering Department, Kyung Hee University, Suwon, South Korea

3 Faculty of Marine Engineering, Vietnam Maritime University, Haiphong, Vietnam

*Address all correspondence to: bapd.vck@vimaru.edu.vn

References

[1] Ba PD, Lee SG, Back S, Kim J, Lee MK. Balancing and translation control of a ball Segway that a human can ride. In: 2016 16th International Conference on Control, Automation and Systems (ICCAS). 2016. pp. 477-480

[2] Kumaga M, Ochiai T. Development of a robot balanced on a ball—Application of passive motion to transport. In: 2009 IEEE International Conference on Robotics and Automation (ICRA'09). 2009. pp. 4106-4111

[3] Lauwers TB, Kantor GA, Hollis RL. A dynamically stable single-wheeled mobile robot with inverse mouse-ball drive. In: Proceedings 2006 IEEE International Conference on Robotics and Automation, 2006 (ICRA 2006). 2006. pp. 2884-2889

[4] Ching-Chih T, Ming-Han J, Cheng-Kai C, Ching-Wen L, Siang-Jyun C. Self-balancing and position control using multi-loop approach for ball robots. In: 2010 International Conference on System Science and Engineering (ICSSE). 2010. pp. 251-256

[5] Kantor G, Hollis H, Nagarajan U. The Ballbot: An omnidirectional balancing mobile robot. The International Journal of Robotics Research. 2014;**33**(6):917-930

[6] Sukvichai K, Parnichkun M. Double-level ball-riding robot balancing: From system design, modeling, controller synthesis, to performance evaluation. Mechatronics. 2014;**24**(5):519-532

[7] Lotfiani A, Keshmiri M, Danesh M. Dynamic analysis and control synthesis of a spherical wheeled robot (Ballbot). In: 2013 First RSI/ISM International Conference on Robotics and Mechatronics (ICRoM). 2013. pp. 481-486

[8] Zabihi H, Talebi HA, Suratgar AA. Open-loop trajectory planning and nonlinear control for underactuated spherical wheel mobile robot (Ballbot). In: 2016 24th Iranian Conference on Electrical Engineering (ICEE). 2016. pp. 549-554

[9] Chih-Hui C, Wen-Ru T. Design and implementation of an omnidirectional spherical mobile platform. IEEE Transactions on Industrial Electronics. 2015;**62**(3):1619-1628

[10] Ching-Chih T, Chang Hsuan C, Feng-Chun T, Kao-Shing H. Adaptive RFWCMAC cooperative formation control for multiple Ballbots incorporated with coupling dynamics. In: 2015 International Conference on Informative and Cybernetics for Computational Social Systems (ICCSS). 2015. pp. 59-65

[11] Denker A, Ohnishi K. Robust tracking control of mechatronic arms. IEEE/ASME Transactions on Mechatronics. 1996;**1**(2):181-188

[12] Wang W, Yi J, Zhao D, Liu D. Design of a stable sliding-mode controller for a class of second-order underactuated systems. IEE Proceedings - Control Theory and Applications. 2004;**151**(6):683-690

[13] Yan M-X, Jing Y-W, He Y-G, Sun P. Adaptive sliding mode controller for a class of second-order underactuated systems. In: 2009 Chinese Control and Decision Conference (CCDC'09). 2009. pp. 2782-2786

[14] Tuan LA, Lee S-GL, Moo S-C. Partial feedback linearization and sliding mode techniques for 2D crane control. Transactions of the Institute of Measurement and Control. 2014;**36**(I): 78-87

[15] Tuan LA, Lee S-G, Nho LC, Kim DH. Model reference adaptive sliding mode control for three dimensional overhead cranes. International Journal of Precision Engineering and Manufacturing. 2013; **14**(8):1329-1338

[16] Tuan LA, Kim J-J, Lee S-G, Lim T-G, Nho LC. Second-order sliding mode control of a 3D overhead crane with uncertain system parameters. International Journal of Precision Engineering and Manufacturing. 2014; **15**(5):811-819

[17] Tuan LA, Lee S-G. Sliding mode controls of double-pendulum crane systems. Journal of Mechanical Science and Technology. 2013;**27**(6):1863-1873

[18] Karkoub MA, Zribi M. Robust control schemes for an overhead crane. Journal of Vibration and Control. 2001; 7(3):395-416

[19] Almutairi NB, Zribi M. Sliding mode control of a three-dimensional overhead crane. Journal of Vibration and Control. 2009;**15**(11):1679-1730

[20] Ching-Wen L, Ching-Chih T, Yi Yu L, Cheng-Kai C. Dynamic modeling and sliding-mode control of a Ball robot with inverse mouse-ball drive. In: SICE Annual Conference 2008. 2008. pp. 2951-2955

[21] Cheng-Kai C, Ching-Chih T. Intelligent backstepping sliding-mode control using recurrent interval type 2 fuzzy neural networks for a ball robot with a four-motor inverse-mouse ball drive. In: 2012 Proceedings of SICE Annual Conference (SICE). 2012. pp. 1281-1286

[22] Weiss A, Langlois RG, Hayes MJD. The effects of dual row omnidirectional wheels on the kinematics of the Atlas spherical motion platform. Mechanism and Machine Theory. 2009;**44**(2):349-358

[23] Slotine JJE, Li W. Applied Nonlinear Control. The United States of America: Prentice-Hall, Inc.; 1991. p. 461

[24] van der Blonk K. Modeling and control of a ball-balancing robot [internship & master thesis]. Electrical Engineering, Mathematics and Computer Science, University of Twente, Alten; 2014

Chapter 6

PID Gain Tuning for Robust Control of PMDC Motor for External Disturbance Rejection with Constrained Motor Parameter Variations through H∞

Prasanth Venktareddy, Prashanth Narayanappa Anand and Prakasha Pundareekane Kanchappa

Abstract

This chapter describes the controller modeling for PID gain tuning against the external disturbances with constrained internal parameter variation of the PMDC motor based on an optimization technique of H-infinity. To fit the goals in the H infinite framework, auto tuning of the PID controller gains is used. Different performance goals for tracking are preset as design objectives. Researchers in literature have presented many Robust Control techniques for motor control applications. Methods like back-stepping algorithms, fuzzy and neural based control systems, model predictive control and SMC (Sliding Mode Control) are available in literature. In this chapter, SC (speed control) of PMDC-motor is addressed with variations in outer load disturbances and internal variations of the system parameters for a particular application. C-PID (conventional PID controllers) is preferred, and equivalent robustness characteristics are established using the H-infinity development procedures. The optimization effort is to get simultaneous fast-tracking response and better disturbance rejection.

Keywords: H-infinity, C-PID, PMDC motor, robust control, disturbance rejection

1. Introduction

This necessity of controller design modeling for proportional integral and derivative (PID) gain tuning against the external disturbances with constrained internal parameter variation of the permanent magnet direct current (PMDC) motor based on an optimization technique of H-infinity is highly recommended. Here, PID controller with auto tuning of gains are used to match the goals in H infinity framework. Different performance goals for tracking are preset as design objectives. The speed control of PMDC-motor has so far attracted the attention of the researchers in recent times and many approaches and improvements have been proposed for PMDC motor

IntechOpen

drives. PMDC motors are used in electrical equipment, computer peripherals, and manipulators because of their precise speed control capabilities. C-PID controllers have been in use since several years for different applications such as motor-control. The classical tuning methods of PID controller, as well as response method of Zeigler-Nichols frequency considers the system in the mode of oscillation to analyze the tuning procedure [1]. Since the manual tuning of PID controller is not user friendly, as it tends to be a tedious process though being simple in structure.

We know that PID controller fails to address instant tracking / regulation to get robustness against disturbance rejection. The majority of the time, industrial controllers are not properly tuned. Traditionally, motor control applications have better performed in control execution for specific working conditions. Controller parameters can be tuned for exact working conditions with an underlying assumption that the conditions are ordered and defined. Basically, working conditions according to the framework that are prone to variations leads to undesirable results if the variations are not accurately introduced. As these controllers do not guarantees the robustness to inner and outer disturbances, the outer disturbances and the system parameter variations have huge impact on performance and degradation in the applications of motor control.

Researchers in literature have presented many Robust Control techniques for motor control applications. Methods like back-stepping algorithms, fuzzy and neural based control systems, model predictive control and sliding mode control (SMC) are available in literature. The SMC based methodology stimulates the chattering phenomenon due to switching function of the inherent discontinuity. In Eker [2], the authors have demonstrated SMC for various applications of PMDC motor control. In Mamani et al. [3] and Corradini et al. [4], semi-SMC, adaptive SMC and boundary layer control (BLC) have been introduced as the development to classical style of Controlling the SMC and reducing chattering impact. Until now, most of the heuristic engineering methods have been developed to achieve optimal tuning of PID.

PID control with GA optimization based for the DC motor is implemented in Pal et al. [5] and can be stimulated by the aid of normal development methods. These methods have proven that the goal functions have degraded [6].

The PSO (Parasitic Swarm Optimization) is considered as the method of ideal structure to control of BLDC motor by PID control in Nasri et al. [7]. The major advantages being ease of implementation and computational complexity, which facilitates meeting constraints for some of the parameters apart from exhibiting fast convergence.

Position and speed control applications have been implemented using H-infinity-based control [8] providing better robustness in performance against disturbances, making this as an attractive alternative. To develop the Robust Controller, the NN based H-infinity controller was presented in Premkumar et al. [9] that introduces the DC motor H-infinity controller and has addressed parameter uncertainties.

In this chapter, SC application of PMDC-motor is addressed with variations in outer load disturbances and internal variations of the system parameters. Controller is also opted as C-PID and equivalent robustness characteristics are established using the H-infinity development procedures. The optimization effort is to get simultaneous fast-tracking response with better disturbance rejection.

2. PMDC motor model

Using supply current, motor speed and supply voltage relationship, we have the PMDC motor mathematical model as:

Parameter	Nominal value	Variation (tolerable limit)
L	0.00063 H	41%
R	2.07 Ohm	40%
Viscous-friction (K_f)	0.000049 Nms rad^{-1}	51%
EMF constant (K_b)	0.053 Vs rad^{-1}	[0.013 to 0.1]
Armature-constant (K_m)	0.053 NmA^{-1}	[0.012 to 0.1]

Table 1.
The parameters of motor and their acceptable limits of variation.

$$\frac{di}{dt} = -\frac{R}{L}i - \frac{k_b}{L}\omega + \frac{1}{L}u \tag{1}$$

$$\frac{d\omega}{dt} = \frac{1}{L}k_m i - \frac{1}{J}k_f\omega \tag{2}$$

Here, R and L are resistance and inductance of motor coil, k_b is denoted as the constant of back Electromotive force, J is rotor inertia and k_f is friction constant.

The representation of SS (state space) matrix of similar set of the equations is given by:

$$\frac{d\omega}{dt}\begin{bmatrix} i \\ \omega \end{bmatrix} = \begin{bmatrix} -\frac{R}{L} & -\frac{K_b}{L} \\ \frac{K_m}{J} & -\frac{K_f}{J} \end{bmatrix}\begin{bmatrix} i \\ \omega \end{bmatrix} + \begin{bmatrix} \frac{1}{L} \\ 0 \end{bmatrix} \tag{3}$$

$$y(t) = [0\ 1]\begin{bmatrix} i \\ \omega \end{bmatrix} + [0]u(t) \tag{4}$$

The overall transfer function for the speed and input voltage, is expressed as:

$$\frac{\omega(s)}{u(s)} = \frac{K_m}{JLS^2 + (JR + LK_f)s + K_mK_b + RK_f} \tag{5}$$

The transfer function of PMDC motor model is considered for a specified output speed and input voltage provided. The unknown motor parameters are prone to the variations due to aging of motor and its wear and tear. The controller must be built to offer resilience for each motor parameter within acceptable variation limits. The parameters of nominal motor are chosen from the standard motor of Maxon-RE35. The variation limits and motors parameters are given in **Table 1**. The proposed controller design for disturbance rejection in this work is based on the presumption that the PMDC motor internal parameter variations are within tolerable limits as specified for Maxon-RE35. The proposed Robust Controller design is discussed in subsequent section.

3. H∞ based robust controller design for disturbance rejection

The proposed controller design model should meet the desired specifications within bounds of internal parameter variations. Figure below shows the variation in the gain of PMDC plant against variations of internal parameters. Similar variations are represented in the domain of frequency via bode plot, that are shown in **Figures 1** and **2**. Henceforth, PID controller is designed to address variation in the model and provide desired specifications for closed loop performance.

This chapter discusses the target specifications for better disturbance rejection and simultaneously set the point tracking. Here, we assume that the control bandwidth, exhibits increasing slope at the crossover frequency compared to disturbance rejection, allowing better gain within the bandwidth. The higher slope is described as lesser phase margin resulting in acceptable overshoot in response to set point. In order to match competing requirements of the tracking rejection and disturbance, the PID controller with 2-DOF (Degree of Freedom) is used as transfer function.

$$u = K_P(br - y) + \frac{K_i}{s}(r - y) + \frac{K_D s}{1 + T_f s}(cr - y) \tag{6}$$

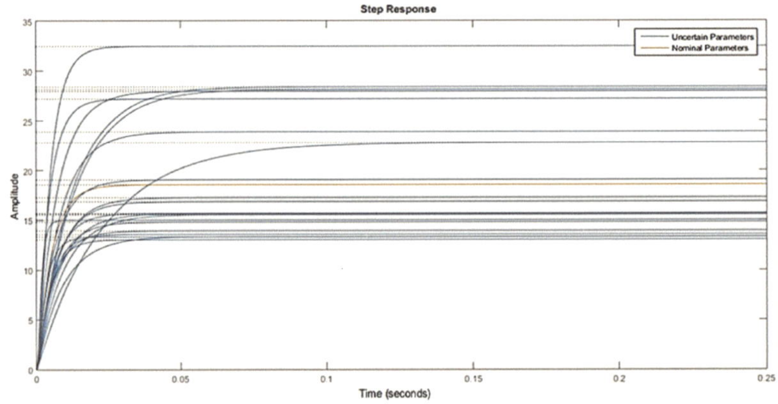

Figure 1.
The variations in PMDC gain of motor plant for uncertain and nominal internal parameter variations.

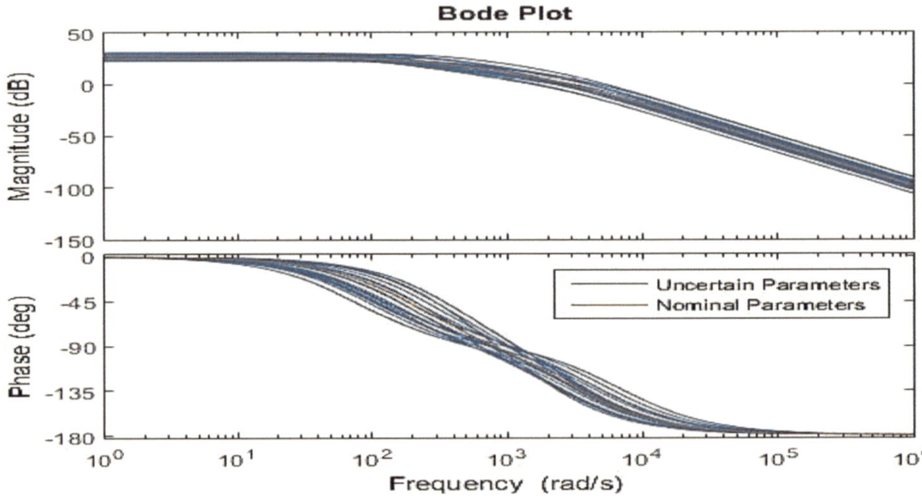

Figure 2.
Bode plot of plant model subjected to the internal parameter variations.

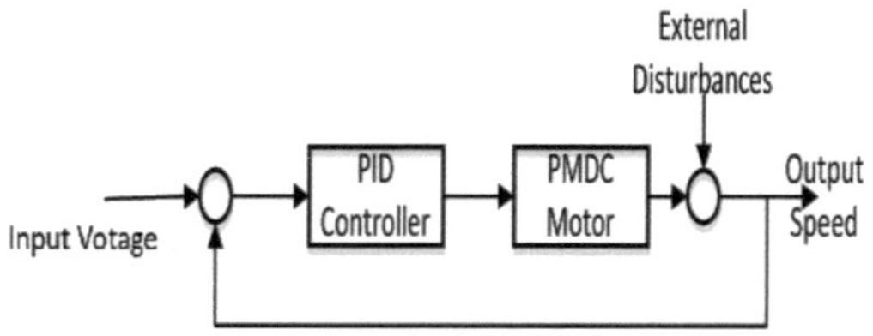

Figure 3.
Block diagram with external disturbances (load variations).

2-DOF (Degree of Freedom) PID controllers contain weighing coefficients associated with derivative and proportional terms. These weighing coefficients facilitate effective disturbance rejection restraining maximization of the over-shoot under optimal conditions. The PID controller of 2-DOF works better in moderating the changes arising in the reference signal or control-signal.

The close loop control system is considered as shown in **Figure 3**. Here, the external disturbances such as load variations on motor shafts and its torque variations are added that influences on the motor parametric variations.

4. The performance goals for the optimization of H∞

The 2-DOF for controller can be tuned by utilizing the approach of H infinity optimization. The open loop gain being a critical indicator of feedback loop behavior, gain of open loop should be more than one in control bandwidth to confirm better DR (Disturbance Rejection) and it must be lower than one outer of control bandwidth to confirm insensitivity to measurement noise of unmodelled dynamics. The ideal performance terms can be displayed regarding execution objectives. To accomplish a decent disturbance rejection and tracking, three execution objectives/limitations are forced on the tuning of controller gain (see discussion in previous section), which is as follows:

i. "Tracking" – used to identify the RT (response time) to step input

ii. "ML (Minimum loop) gain" – used to recognize loop gain before the frequency of crossover.

iii. "ML (Maximum loop) gain" – used to to identify the control bandwidth at higher frequencies.

The CGs (Controller-gains) must be tuned with constraint i.e., function cost, connected with every specification subjected to minimization in H-infinity framework.

4.1 Tracking as an execution objective

The frequency domain specification for monitoring between output and input is described in this performance target. This frequency domain constraint indicates the most extreme relative error as a FF (frequency function). The ME (maximum error) is given by:

$$error_{max} = \frac{(error_{peak})s + \omega_c(error_{dc})}{s + \omega_c} \tag{7}$$

where, ω_c denotes cut-off frequency.

The scalar function $f(x)$ describes the tracking goal, where x is denoted as a tuneable vector of entire parameters in the system. The target optimization (TO) is to modify the parameter function $f(x)$, which is optimized. The scalar function for tracking case is described using $f(x)$:

$$f(x) = \left\| \frac{1}{error_{max}} (T(s,x) - I) \right\|_\infty \tag{8}$$

where, $T(s, x)$ is symbolized as closed loop transfer function between input and output.

4.2 Min-LG (minimum) as execution limitation

The minimal gain on the open loop frequency response at specified frequencies is limited by this performance goal. The frequency dependent minimum gain constraint in turn gives the inverse sensitivity function of minimum gain limit. The min constraint gain characterizes the capacity of scalar function $f(x)$ whereas the advancement procedure attempts to drive lower value of $f(x)$. The capacity of scalar $f(x)$ is defined as:

$$f(x) = \left\| W_S(D^{-1}SD) \right\|_\infty \tag{9}$$

where, W_S is denoted as Min-LG profile and S is defined as sensitivity function.

4.3 Max loop gain as execution limitation

This execution goal aims at highest gain of open loop at the determined frequencies in the given framework. The Max-LG can be characterized as the frequency domain element. This type of constraint restrains upper limit on the corresponding sensitivity. The maximum loop gain determines the scalar function $f(x)$ given by:

$$f(x) = \left\| W_T(D^{-1}TD) \right\|_\infty \tag{10}$$

where, W_T is symbolized as the reciprocal profile of the Max-LG. T is symbolized as the function of complementary sensitivity.

5. Performance goal description

The three performance goals for the H-infinity minimization such as tracking, maximum LG and minimum LG. The limits/range for these goals is expected as represented in **Figures 4–6** as:

- Tracking better than 2 sec

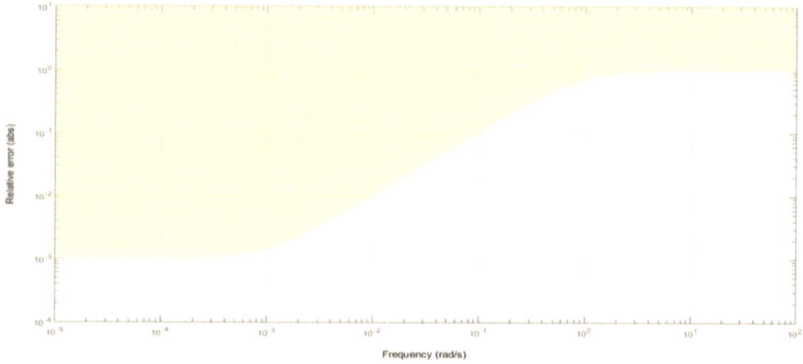

Figure 4.
Performance goal 1: The response for desired tracking.

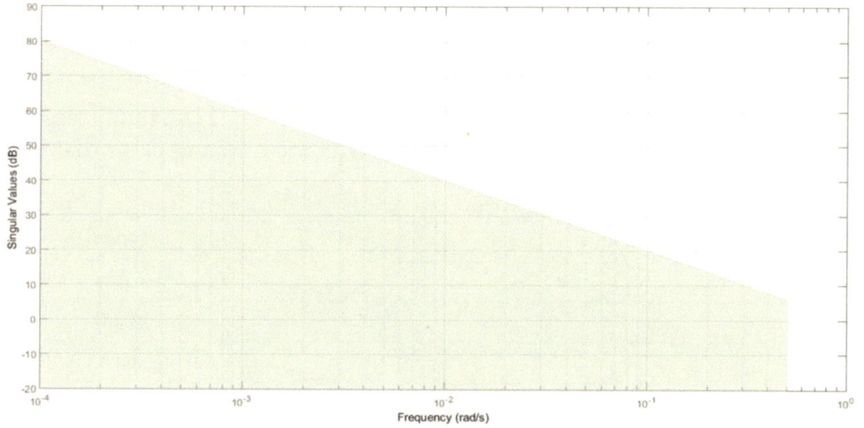

Figure 5.
Performance goal 2: The desired min-LG.

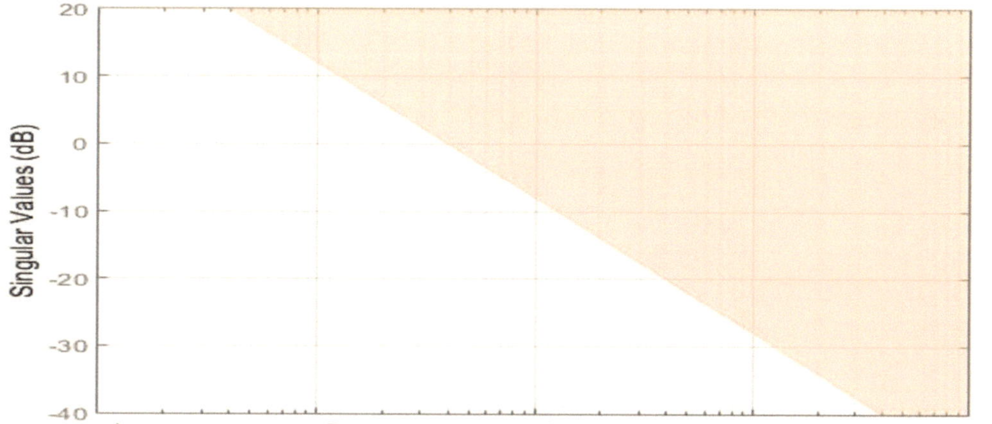

Figure 6.
Performance goal 3: The desired max-LG.

- Min-LG: to be higher below 0.5 rad/s.
- Max-LG: to be less beyond 4 rad/s and the roll off with at least 20db/decade.

6. Simulations and results

The motor model transfer function as described in Eq. (5) along with the constraints from Eqs. (6)–(10) is simulated using MATLAB Simulink.

To model the uncertainty of plant parameters, the specified motor parameters can be used as variables. The PID controller of 2-DOF is characterized to have tunable gains. The analysis point is where disturbance torques are calculated and disturbance sensitiveness is determined. The performance goals as examined in Section 5 are selected and H∞ minimization of the scalar function relating to the described performance goals is evolved in this work. This results in optimal tuning of controller gains and is tabulated as shown in **Table 2**.

The outcomes are presented in the following figures.

The **Figure 7** represents response of TE (tracking error) of the tuned function of closed loop transfer function. This plot represents the tracking error that is

Parameter	Value
K_p	0.034
T_f	28.447
K_d	−0.942
K_i	0.341

Table 2.
The resulting tuned parameters of 2-DOF PID from simulation for optimization.

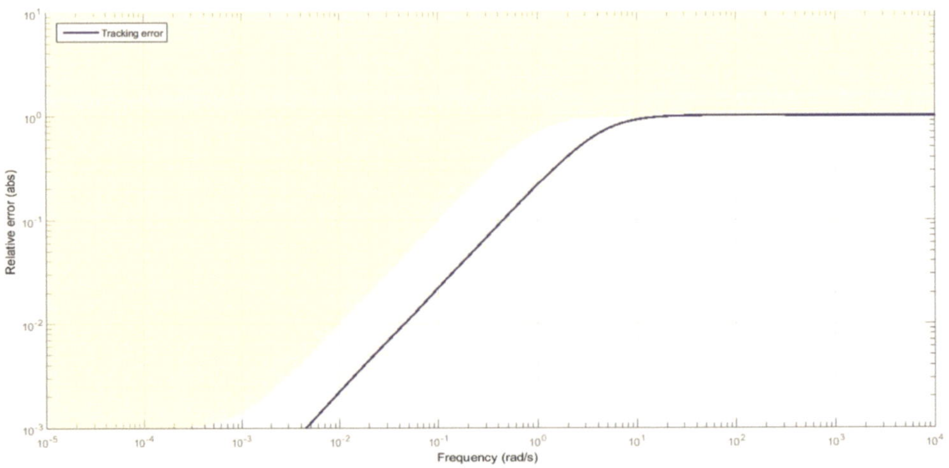

Figure 7.
The performance goal 1: Desired v/s achieved.

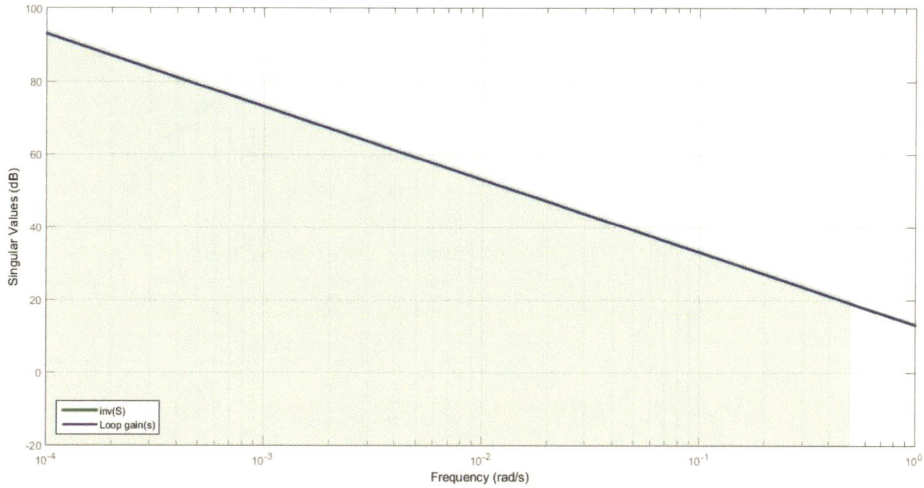

Figure 8.
The performance goal 2: Desired v/s achieved.

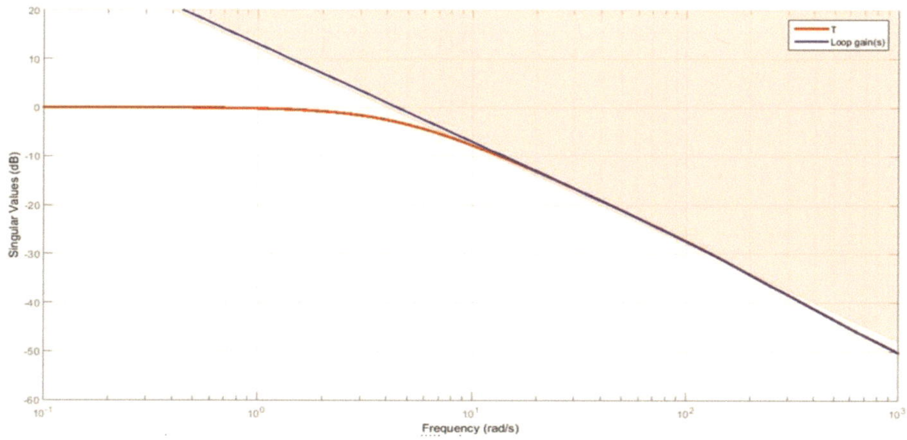

Figure 9.
The performance goal 3: Desired v/s achieved.

accomplished over all ranges of frequencies below the specified margin. Correspondingly, the max-LG and achieved min-loop v/s ideal values are plotted in **Figures 8** and **9**.

Figure 8 shows the achieved transient duration of proposed model is much less that desired transient duration. **Figures 10** and **11** demonstrates tracking performance of the proposed model for any arbitrary input. In all cases, the set point of tracking can be accomplished well within the prescribed 2 seconds limitations in system response. From these plots, we infer that simulation of the proposed model guarantees motor internal parameters to remain within tolerable limits of variations for any applied arbitrary external disturbances. Further, the proposed model is tested for simulation with 30 different randomly chosen external disturbances.

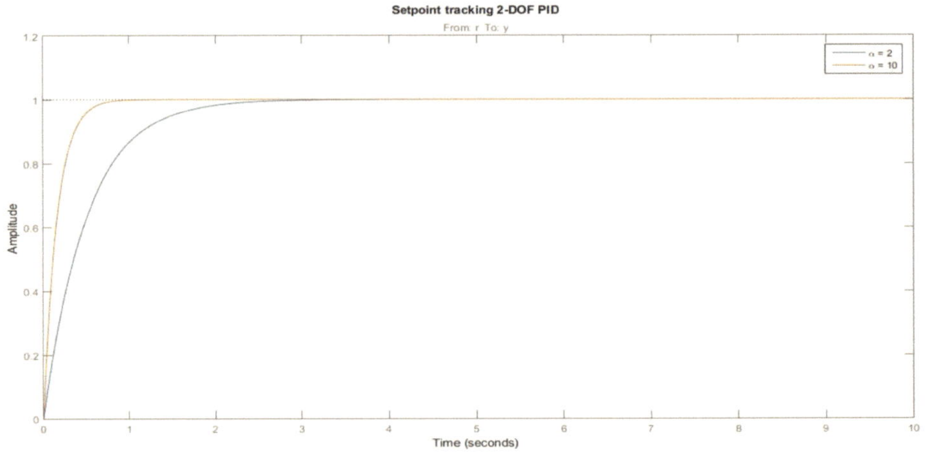

Figure 10.
The performance of tracking with 2-DOF PID controller.

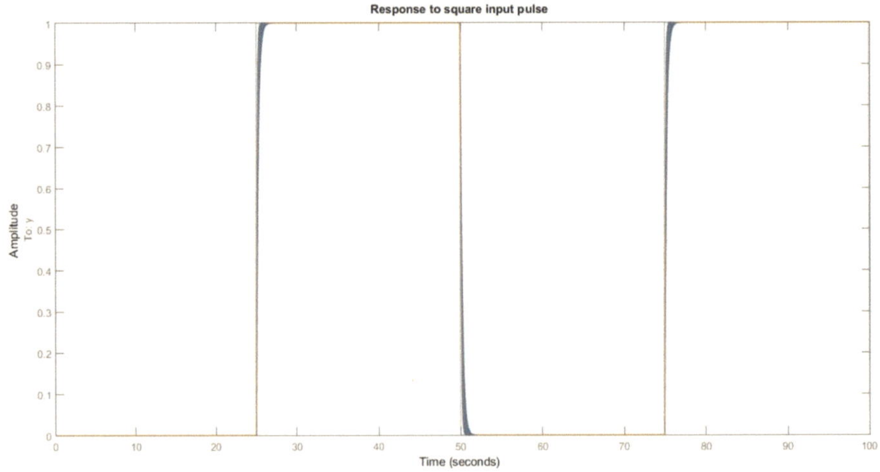

Figure 11.
The performance of speed control to the pulse command (tracking better than the 2 sec).

It is significant to note that achieved performances are far superior to the focused objectives. Additionally, H_∞ advancement of targeted goals outperforms over expected performance. Out of several performance goals specified for PMDC motors, our work presents simulation results for only three objectives. However, other goals remain to be explored.

Figure 12 represents the disturbance rejection achieved for two different values of proportional and derivative coefficients. The response for randomly chosen load disturbances (30 arbitrary signals) with implicit effect on internal parameters is presented in **Figure 13**. It can be observed that the response for the proposed robust model for controller is well within the accepted bounds. It is worth to observe that the disturbances attenuate well below the prescribed limit of 2 sec.

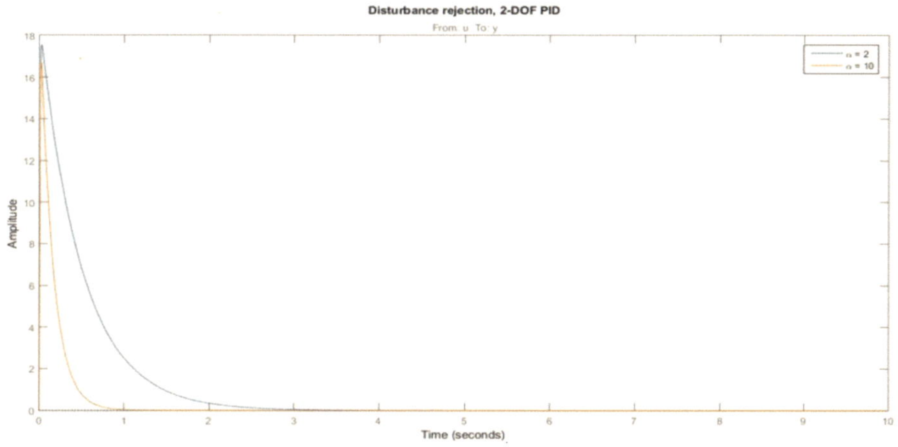

Figure 12.
Disturbance rejection with the controller of 2-DOF PID.

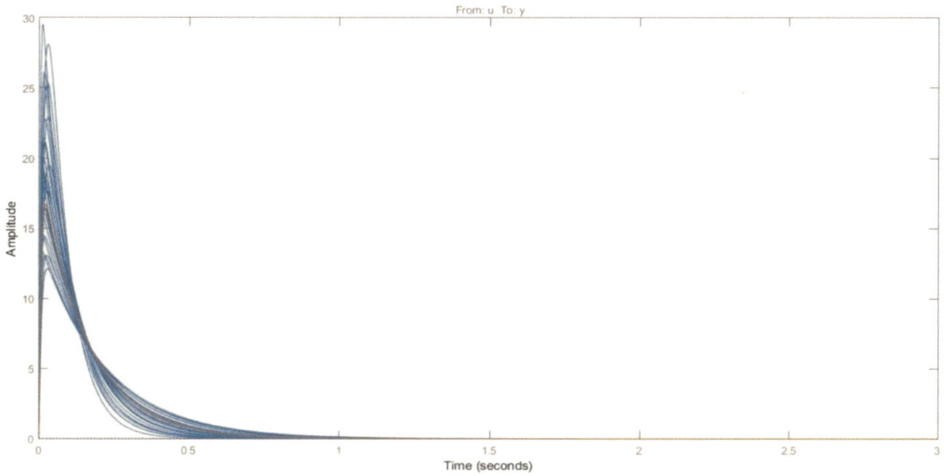

Figure 13.
The response to 30 random disturbances with implicit parameter variations.

7. Conclusion

This work introduces the MO-optimization (Multi Objective) with three performance goals in the structure of H∞ for tuning parameters of PID controller for PMDC motor. The two simultaneous competing needs of disturbance rejection along with input tracking are achieved with Robust Controller design with 2-DOF over disturbance rejection using H∞ framework. Hence, it is established that the PID controller can be tuned, and corresponding gains can be achieved for a specified constraints/ performance capability. Further, H∞ provides a better and acceptable framework for optimization.

Author details

Prasanth Venktareddy[1*], Prashanth Narayanappa Anand[2] and
Prakasha Pundareekane Kanchappa[3]

1 Nitte Meenakshi Institute of Technology, Bangalore, India

2 BMS Institute of Technology and Management, Bangalore, India

3 Don Bosco Institute of Technology, Bangalore, India

*Address all correspondence to: prasanth.v@nmit.ac.in

References

[1] Soni R, Singh D, Pandey P, Sharma P. Simulation of optimal speed control for a DC motor using conventional PID controller and fuzzy logic controller. International Journal of Information and Computation Technology. 2013;**3**: 181-188

[2] Eker I. Sliding mode control with PID sliding surface and experimental application to an electromechanical plant. ISA Transactions. 2006;**45**(1): 109-118. DOI: 10.1016/S0019-0578(07) 60070-6

[3] Mamani G, Becedas J, Batlle VF. Robust position control of a DC motor by sliding mode. IFIP Advances in Information and Communication Technology. 2010;**314**:495-504. DOI: 10.1007/978-3-642-11628-5_55

[4] Corradini ML, Ippoliti G, Longhi S, Orlando G. A quasi-sliding mode approach for robust control and speed estimation of PM synchronous motors. IEEE Transactions on Industrial Electronics. 2012;**59**(2):1096-1104. DOI: 10.1109/TIE.2011.2158035

[5] Pal P, Dey R, Pal P, Biswas RK, Bhakta S. Optimal PID controller Design for Speed Control of a separately excited DC motor: A firefly based optimization approach. International Journal of Soft Computing, Mathematics and Control. 2015;**4**(4):39-48. DOI: 10.2139/ ssrn.3517152

[6] He J-B, Wang Q-G, Lee T-H. PI/PID controller tuning via LQR approach. Chemical Engineering Science. 2000;**55**: 2429-2439. DOI: 10.1016/S0009-2509 (99)00512-6

[7] Nasri M, Nezamabadi-pour H, Maghfoori M. A PSO-based optimum design of PID controller for a linear brushless DC motor. Proceedings of World Academy of Science, Engineering And Technology. 2007;**20**:2

[8] Bautista-Quintero R, Pont MJ. Implementation of H-infinity control algorithms for sensor-constrained mechatronic systems using low-cost microcontrollers. IEEE Transactions on Industrial Informatics. 2008;**4**(3): 175-184. DOI: 10.1109/TII.2008.2002703

[9] Premkumar M, Ram JVR, Jazadamba KPVPB, Rammohan K. Design and simulation of ARTD controller for chemical process industry PMDC motor. Proceedings of the 4th International Conference on Electrical Energy Systems. 2018:454-458. DOI: 10.1109/ICEES.2018.8443185

Chapter 7

Role of Uncertainty in Model Development and Control Design for a Manufacturing Process

Rongfei Li and Francis F. Assadian

Abstract

The use of robotic technology has drastically increased in manufacturing in the twenty-first century. But by utilizing their sensory cues, humans still outperform machines, especially in the micro scale manufacturing, which requires high-precision robot manipulators. These sensory cues naturally compensate for high level of uncertainties that exist in the manufacturing environment. Uncertainties in performing manufacturing tasks may come from measurement noise, model inaccuracy, joint compliance (e.g., elasticity) etc. Although advanced metrology sensors and high-precision microprocessors, which are utilized in nowadays robots, have compensated for many structural and dynamic errors in robot positioning, but a well-designed control algorithm still works as a comparable and cheaper alternative to reduce uncertainties in automated manufacturing. Our work illustrates that a multi-robot control system can reduce various uncertainties to a great amount.

Keywords: uncertainty, modeling, feedback control design, automated manufacturing, robot arm system

1. Introduction

It is believed that the rapid emergence of Robotic technology in industry, and specifically in manufacturing, in the century, will have positive impacts in many aspects of our lives. We have already seen many applications of this technology in macro scale, such as pick and place task [1]. However, there are still applications where humans outperform machines, especially in the micro scale manufacturing, which requires high-precision robot manipulators.

Accurate positioning of robot arms is very important in automated manufacturing field. Over past several decades, we have seen great strides in the technology for accurate positioning robots. We have seen researchers have tried to implement add-on features such as real-time microprocessors, high precision motors, zero backlash gear set, advanced metrology sensors and so on in today's robots. Indeed, they have compensated many structural and dynamic errors in robot positioning [2]. However, those add-on features are usually very expensive and unnecessarily increase the cost during the manufacturing process. Robotic systems that employ a well-designed sensor-based control strategies can reduce the cost and simultaneously obtain robustness against disturbances and imprecisions from sensors or modeling.

The process of fastening and unfastening a screw is a mundane but a challenging task in the automated manufacturing. We found recent research on this topic only focuses on how robots should generate push/pull force on a driver [3]. The axial forces and torques are first measured through sensors and then controlled to imitate human approach of fastening and unfastening by applying similar amount of axial forces and torques. This approach only considers tactile sensing, however, human beings, also use the information from visual sensing to help with this task. To replicate visual sensing in robots, for example, a camera system could be utilized to make sure a tool is at the right pose (correct orientation and location where head of bolt and tail of driver coincide). A visual system that can provide an accurate and repeatable positional tracking of the tool becomes significantly important and useful not only in this type of an application but also in many other applications of the automated manufacturing [4].

In this work, we have designed a multi-robotic control system that simulates the positioning process for fastening and unfastening applications and have examined its robustness against various uncertainties, which may occur, in this process. This control system is a visual servoing system where a camera is mounted on a robot arm manipulator and provides vision data for the motion control of a second robot manipulator with a tool. Both the Position-Based Visual Servoing (PBVS) and the Image-Based Visual Servoing (IBVS) systems have been thoroughly investigated in [5–8]. However, in this related work, in the visual servoing domain, the development of the outer-loop controller is usually achieved with the PID controller, or its simplified variations based only on a kinematic model of camera [5, 8]. One improvement in this work is to use Youla robust control design technique [9] that includes both kinematics and dynamics in the model development stage. The increase in the model fidelity for the control design can positively influence the precision of the feature estimation and the control system stability for the high-speed tasks. Benefits of our design are discussed in more details in the following sections.

Position control algorithms for both the visual and the tool manipulation systems are discussed in this Chapter. Especially, a combination of a feedforward and a feedback control architecture has been designed for the tool manipulation system, which enables the tool to move fast to a desired location with a high precision in its final pose. Simulation results for the Single Input Single Output (SISO) case in various scenarios are presented and furthermore, the robustness to various noise sources in this manufacturing process are examined.

2. Literature review

We have seen many efforts been made to improve the positioning accuracy of robotic systems in the past few decades. An effective way to reduce the amount of inaccuracy is to measure it with sensors and compensate is through feedback control loop. Many metrology techniques have been investigated and applied for different kinds of data capturing. Among them, three methods have gained the most popularity in the recent research, namely, vision-based methods, tactile-based methods, and vision-tactile integrated methods. In this section, we will briefly review those approaches and their applications.

2.1 Vision-based methods

The vision-based methods have been widely developed in the recent years and used to determine position and orientation of target objects in robotic systems. Zhu et al. discussed Abbe errors in a 2D vision system for robotic drilling. Four laser displacement sensors were used to improve the accuracy of the vison-based measurement system [10]. Liu et al. proposed a visual servoing method for positioning in aircraft digital assembly [11, 12]. With the measurements from two CCD cameras and four distances sensors, the proposed method can accurately align the positioner's ball-socket with the ball-head fixed on the aircraft structures in a finite time.

In addition to mentioned applications, we have seen many contributions to the vision-based methods in robotic manipulation. However, most of those researchers focused on success rate of grasping on end-effector without enough analysis on the positioning accuracy. Du et al. published a study for the robotic grasp detection by visually localizing the object and estimating its pose [13]. Avigal et al. proposed a 6-DoFs grasp planning method using fast 3D reconstruction and grasp quality convolutional neural network (CNN) [14]. Wu et al. proposed an end-to-end solution for visual learning [15].

2.2 Tactile-based methods

In addition, with the development of tactile sensors in the last few years, we have seen more and more focus on tactile-based methods in robotic positioning domain. The tactile sensors can show contact states of the end-effector and the object in robotic manipulations. The contact state can be used to determines objects' relative orientations and positions with respective to the gripper. Li et al. designed a tactile sensor of GelSight and generated tactile maps for different poses of a small object in the gripper [16]. He studied the localization and control manipulation for a specific USB connector insertion task. Dong et al. studied the tactile-based insertion task for dense box packing with two GelSlim fingers which are used to estimate object's pose error based on neural network [17]. Furthermore, Hogan et al. developed a tacile-based feedback loop in order to control a dual-palm robotic system for dexterous manipulations [18]. Those tactile-based methods can only realize relative accurate positioning of the tool with the end-effector, but the positioning of the robot manipulator itself is not addressed.

2.3 Vision-tactile integrated methods

Vision sensing can provide more environment information with a wide measurement range, while tactile sensing can provide more detailed information in robotic manipulations. Therefore, the vision–tactile integrated methods came into being. Fazeli et al. proposed a hierarchical learning method for complex manipulation skills with multisensory fusion in seeing and touching [19]. Gregorio et al. developed a manipulation system for automatic electric wires insertion performed by an industrial robot with a camera and tactile sensors implemented on a commercial gripper [20].

According to the analysis, all the integrated sensory applications have achieved accurate robotic manipulation tasks such as insertion and their performances have been verified in experiments. However, the error space in those references is usually small and none of them has considered all the translational and rotational errors in 6 DoFs. Moreover, tactile-based or vision-tactile integrated methods will increase expense of massive manufacturing because tactile sensors are more expensive to purchase and maintain compared to visual sensors. Based on that, our work is to explore the capability of the vison-based methods and design a new method to improve the accuracy of positioning in the robot manipulation system.

3. Technical background

In the first part of this section, we will briefly discuss the algorithm of the traditional Image based visual servoing (IBVS) and its drawbacks. In the second part, we will review sources of all possible uncertainties occurred in automated manufacturing and existing approaches for reducing them.

3.1 A brief overview of the classical IBVS architecture

The control of the visual system, the eye-in-hand camera configuration [21], is discussed in this section. The general visual servoing problem can be divided into two categories: PBVS and IBVS [5]. This work focuses on the IBVS structure. **Figure 1** shows the control block diagram for a classical IBVS architecture. In **Figure 1**, $\boldsymbol{s} = [\boldsymbol{u}, \boldsymbol{v}]^T$ is the image feature position vector, $\boldsymbol{s} \quad = [\boldsymbol{u} \quad , \boldsymbol{v} \quad]^T$ is the target image feature position vector, and their difference $\boldsymbol{e} = \boldsymbol{s} \quad - \boldsymbol{s}$ is the error vector. The IBVS structure is a cascaded control loop with an outer controller and an inner joint controller. The outer controller takes the feature error e as input and generates the positional targets, which are denoted as $\dot{\boldsymbol{q}}$ in the **Figure 1**. The inner controller

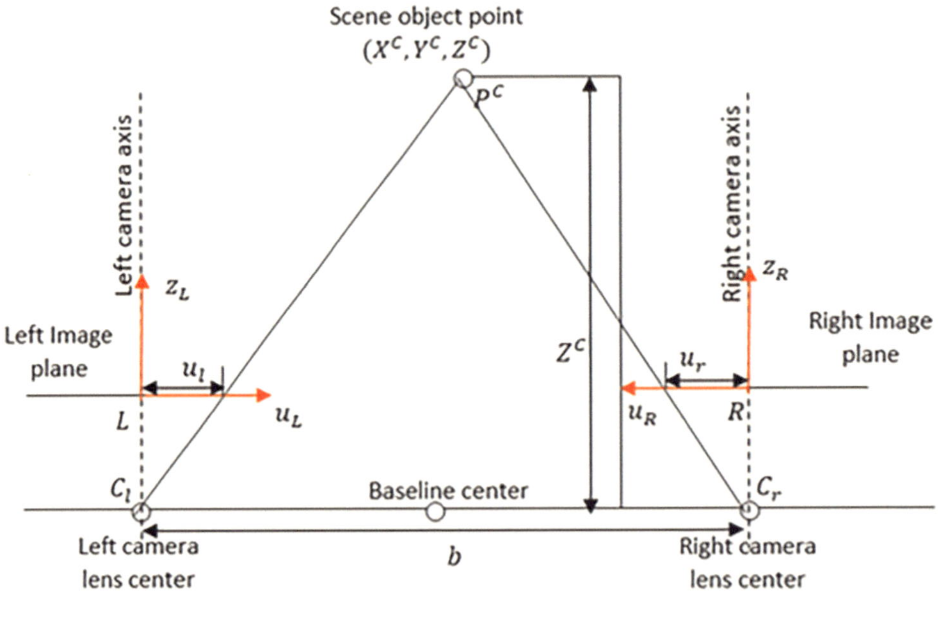

Figure 1.
The block diagram of the classical IBVS control architecture.

stabilizes the joints to generated positional targets from outer controller. L_e is the so-called interaction matrix [5], which is a 2-by-6 matrix, and relates the time derivative of the image feature s to the spatial velocity of the camera V_c, a column vector of six-elements, by the following:

$$\dot{s} = L_e V_c \tag{1}$$

We can design a proportional controller to force the error to exponentially converge to zero, i.e.:

$$\dot{e} = -ke, \;\; k > 0 \tag{2}$$

Suppose the target image feature is a constant; that is $\dot{s} \;\; = 0$, hence, we can derive from Eq. (2):

$$\dot{e} = \dot{s}^* \; - \dot{s} = -L_e V_c \tag{3}$$

From Eqs. (2) and (3), one will be able to obtain:

$$V_c = kL_e^+ e \tag{4}$$

where L_e^+ is the pseudoinverse of L_e. The derivation of the interaction matrix with a monocular camera is further explained next. We assume a point with the three-dimension (3D) coordinates in the camera frame is given as $P^C = [X^C, Y^C, Z^C]$. We further assume a zero-skew coefficient, i.e., $s_C = \mathbf{0}$ in Eq. (2) with a baseline distance $b = \mathbf{0}$, for a monocular camera model, then the image feature coordinate $s = [u, v]^T$ can be expressed as:

$$\begin{bmatrix} u \\ v \end{bmatrix} = \begin{bmatrix} \dfrac{f_u X^C}{Z^C} + u_0 \\ \dfrac{f_v Y^C}{Z^C} + v_0 \end{bmatrix} \tag{5}$$

Taking the time derivative of Eq. (5), we can obtain:

$$\begin{bmatrix} \dot{u} \\ \dot{v} \end{bmatrix} = \begin{bmatrix} \dfrac{f_u\left(\dot{X}^C Z^C - \dot{Z}^C X^C\right)}{(Z^C)^2} \\ \dfrac{f_v\left(\dot{Y}^C Z^C - \dot{Z}^C Y^C\right)}{(Z^C)^2} \end{bmatrix} \tag{6}$$

The rigid body motion of a 3D point in the camera model can be derived as:

$$\dot{P}^C = v^C + \omega^C \times P^C \Longleftrightarrow \begin{cases} \dot{X}^C = v_X^C - \omega_Y^C Z^C + \omega_z^C Y^C \\ \dot{Y}^C = v_Y^C - \omega_Z^C X^C + \omega_Z^C Z^C \\ \dot{Z}^C = v_Z^C - \omega_X^C Y^C + \omega_y^C X^C \end{cases} \tag{7}$$

Substituting (7) into (6), and rearranging the terms, we obtain:

$$\begin{bmatrix} \dot{u} \\ \dot{v} \end{bmatrix} = \begin{bmatrix} \frac{f_u}{Z^C} & 0 & -\frac{u-u_0}{Z^C} & -\frac{(u-u_0)(v-v_0)}{f_v} & \frac{{f_u}^2+(u-u_0)^2}{f_u} & -\frac{f_u(v-v_0)}{f_v} \\ 0 & \frac{f_v}{Z^C} & -\frac{v-v_0}{Z^C} & -\frac{{f_v}^2+(v-v_0)^2}{f_v} & \frac{(u-u_0)(v-v_0)}{f_u} & \frac{f_v(u-u_0)}{f_u} \end{bmatrix} \begin{bmatrix} v_X^C \\ v_Y^C \\ v_Z^C \\ \omega_X^C \\ \omega_Y^C \\ \omega_Z^C \end{bmatrix} \tag{8}$$

Eq. (8) can be simply written as:

$$\dot{s} = L_e V_c = L_e \begin{bmatrix} v^C \\ \omega^C \end{bmatrix} \tag{9}$$

Some drawbacks of the classical IBVS are summarized next. To compute the interaction matrix $\boldsymbol{L_e}$ from Eq. (8), the depth $\boldsymbol{Z^C}$ needs to be estimated. This can be usually approximated as either the depth of the initial position or the depth of the target position or their average value [5]. A careless estimation of the depth may lead to a system instability. In addition, the design of the proportional controller is based on, Eq. (1), the camera kinematic relationships, such that there is no dynamics considered in this model. The kinematic model is sufficient for very slow responding system, however, for faster responses, one has to take into account the manipulator dynamics along with the camera model.

In this work, we propose a new controller algorithm, similar to the classical IBVS structure, where the controller is designed with the complete dynamic and kinematic models of the robot manipulator and the camera. Furthermore, this algorithm does not require any depth estimation, therefore, it will not be necessary to use the interaction matrix. The development of this new algorithm is presented in sections 6 and 7 of this Chapter.

3.2 A brief overview of sources of uncertainties and approaches for reduction

Uncertainties in automated manufacturing can originate from different sources. We can divide these uncertainties into two categories: sensor measurement noise, and dynamic and kinematic modeling errors from both the measurement system and the robot manipulators. This section briefly reviews each uncertainty source including the proposed methods for reducing these uncertainties.

3.2.1 A brief overview of a stereo camera model and its calibration

A camera model (i.e., the pin-hole model [22]) has been adopted in the visual servoing techniques to generate an interaction matrix [5]. The object depth, the distance between a point on the object and the camera center as illustrated in **Figure 2**, needs to be either estimated or approximated by an interaction matrix [5]. One of the methods is to directly measure the depth by a stereo (binocular) camera with the use of two image planes [23].

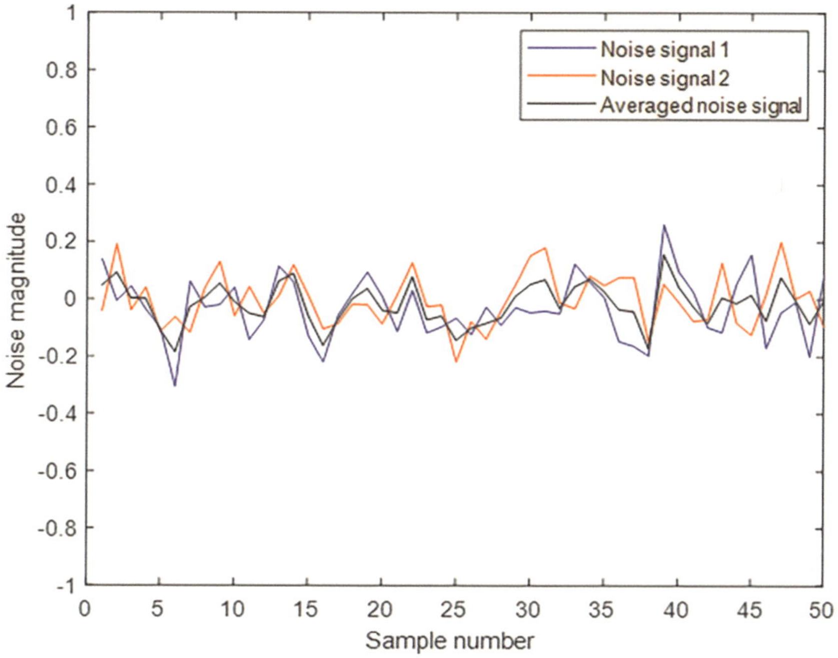

Figure 2.
The projection of a scene object on the stereo camera's image planes.

As shown in **Figure 2**, two identical cameras are separated by a baseline distance b. An object point, $\boldsymbol{P}^C = \left[\boldsymbol{X}^C, \boldsymbol{Y}^C, \boldsymbol{Z}^C\right]^T$, which is measured in the camera frame, at the baseline center, is projected to two parallel virtual image planes, and each plane is located between each optical center ($\boldsymbol{C_l}$ or $\boldsymbol{C_R}$) and the object point $\boldsymbol{P}^C$. The intrinsic camera parameters relate the coordinates of the object point in the camera frame and its corresponding image coordinates $\boldsymbol{p} = (\boldsymbol{u}, \boldsymbol{v})$ on each of the image plane with an exact mathematical relationship. This relationship is given by:

Note: $\boldsymbol{v}$ coordinate on each image plane is not shown in the plot but is measured along the axis that is perpendicular to and point out of the plot.

$$\begin{bmatrix} u_l \\ v_l \\ 1 \end{bmatrix} = \frac{1}{Z^C} \begin{bmatrix} f_u & s_c & u_0 \\ 0 & f_v & v_0 \\ 0 & 0 & 1 \end{bmatrix} \begin{bmatrix} X^C \\ Y^C \\ Z^C \end{bmatrix} - \frac{b}{2Z^C} \begin{bmatrix} f_u \\ 0 \\ 0 \end{bmatrix} \tag{10}$$

$$\begin{bmatrix} u_r \\ v_r \\ 1 \end{bmatrix} = \frac{1}{Z^C} \begin{bmatrix} f_u & s_c & u_0 \\ 0 & f_v & v_0 \\ 0 & 0 & 1 \end{bmatrix} \begin{bmatrix} X^C \\ Y^C \\ Z^C \end{bmatrix} + \frac{b}{2Z^C} \begin{bmatrix} f_u \\ 0 \\ 0 \end{bmatrix} \tag{11}$$

Where, $\boldsymbol{f_u}$ and $\boldsymbol{f_v}$ are the horizontal and the vertical focal lengths, and, $\boldsymbol{s_c}$ is a skew coefficient. In most cases, $\boldsymbol{f_u}$ and $\boldsymbol{f_v}$ are different if the image horizontal and vertical axes are not perpendicular. In order not to have negative pixel coordinates, the origin of the image plane will be usually chosen at the upper left corner instead of the center. $\boldsymbol{u_0}$ and $\boldsymbol{v_0}$ describe the coordinate offsets. The camera model uncertainties may arise from the estimation of those camera intrinsic parameter values. The camera calibration can be used to precisely estimate these values.

The stereo camera calibration has been well studied in [24–27]. As summarized in [24], the calibration method can be divided into two broad categories: the photogrammetric calibration and the self-calibration. In the photogrammetric calibration [25], the camera is calibrated by observing a calibration object whose geometry is well known in the 3D space. These methods are very accurate but require an expensive apparatus and elaborate setups [24]. The self-calibration [24, 26, 27] is performed by finding the equivalences between the captured images of a static scene from different perspectives. Although cheap and flexible, these methods are not always reliable [24]. The author in [24] proposed a new self-calibration technique that observe planar pattern at different orientations and showed improved results.

3.2.2 A brief overview of the robot manipulator model and its calibration

In this work, we consider the elbow manipulators [28] with the spherical wrist in the multi-robot system to move an end-effector freely in 6 degrees of freedoms (dofs). This model of the robot has six links with three for the arms and the other three for the wrist. The robot arms freely move the end effector to any position in the reachable space with 3 dofs while the robot spherical wrists allow the end effector to orient in any directions with another 3 dofs. For the elbow manipulators, a joint is connected between each two adjacent links and there are in total six convolutional joints. The specific industry model of this type is ABB IRB 4600 [29].

A commonly used convention for selecting and generating the reference frames in the robotic applications is the Denavit-Hartenberg convention (or D-H convention) [30]. Suppose each link is attached to a Cartesian coordinate frame, $\boldsymbol{O_iX_iY_iZ_i}$. In this convention, each homogeneous transformation matrix $\boldsymbol{A_i}$ (from frame $\boldsymbol{i-1}$ to frame $\boldsymbol{i}$) can be represented as a product of four basic transformations:

$$A_i = Rot_{z,q_i} Trans_{z,d_i} Trans_{x,a_i} Rot_{x,\alpha_i} = \begin{bmatrix} c_{q_i} & -s_{q_i}c_{\alpha_i} & s_{q_i}s_{\alpha_i} & a_i c_{q_i} \\ s_{q_i} & c_{q_i}c_{\alpha_i} & -c_{q_i}s_{\alpha_i} & a_i s_{q_i} \\ 0 & s_{\alpha_i} & c_{\alpha_i} & d_i \\ 0 & 0 & 0 & 1 \end{bmatrix} \quad (12)$$

Note: $c_{q_i} \equiv \cos(q_i)$, $c_{\alpha_i} \equiv \cos(\alpha_i)$, $s_{q_i} \equiv \sin(q_i)$, $s_{\alpha_i} \equiv \sin(\alpha_i)$.

Where $\boldsymbol{q_i}$, $\boldsymbol{a_i}$, $\boldsymbol{\alpha_i}$ and $\boldsymbol{d_i}$ are parameters of link $\boldsymbol{i}$ and joint $\boldsymbol{i}$, $\boldsymbol{a_i}$ is the link length, $\boldsymbol{q_i}$ is the rotational angle, $\boldsymbol{\alpha_i}$ is the twist angle and $\boldsymbol{d_i}$ is the offset length between the previous $(\boldsymbol{i-1})^{\boldsymbol{th}}$ **and** the current $\boldsymbol{i^{th}}$ robot links. The quantities of each parameter in (12) are calculated based on the steps in [28].

We can generate the transformation matrix from the base frame $\boldsymbol{O_0X_0Y_0Z_0}(\boldsymbol{P^0})$ to the end-effector frame $\boldsymbol{O_6X_6Y_6Z_6}(\boldsymbol{P^6})$:

$$T_6^0 = A_1^0 A_2^1 A_3^2 A_4^3 A_5^4 A_6^5 \quad (13)$$

If any point with respect to the end effector frame $\boldsymbol{P^6}$ is known, we can calculate its coordinate with respect to the base frame $\boldsymbol{P^0}$ as:

$$P^0 = T_6^0 P^6 \quad (14)$$

In addition, the transformation from the base frame $\boldsymbol{P^0}$ to the end-effector frame $\boldsymbol{P^6}$ can be derived:

$$T_0^6 = \left(T_6^0\right)^{-1} \tag{15}$$

which is used to generate the image coordinates of a point captured by a camera with its center attached to the end effector, from the 3D coordinates of a point in the base frame.

Eq. (13) shows that the position of the end-effector P^{end} (where P^{end} is the origin of the end-effector frame $\boldsymbol{P}^6$) is a function of all the joint angles $\boldsymbol{q} = [\boldsymbol{q}_i | \boldsymbol{i} \in \mathbf{1, 2, 3, 4, 5, 6}]$ and the parameters $\boldsymbol{Pa} = [\boldsymbol{a}_i, \boldsymbol{\alpha}_i, \boldsymbol{d}_i | \boldsymbol{i} \in \mathbf{1, 2, 3, 4, 5, 6}]$:

$$P^{end} = \mathcal{F}(q, Pa) \tag{16}$$

Eq. (16) describes the forward kinematic model of the robot manipulator, which could be utilized to calculate the position of the end effector from the joint angles and the robot parameters. The inverse process is called the inverse kinematic, which the joint angles can be computed from the position and the parameters. The estimation of the robot parameters $\boldsymbol{Pa}$ determines the accuracy of the kinematic models of the robot manipulators.

The paper [31] provides a good summary of the current robot calibration methods. The author of [32] states that over 90% of the position errors are due to the errors in the robot zero position (the kinematic parameter errors). As a result, most researchers focus on the kinematic robot calibration (or level 2 calibration [31]) to enhance the robot absolute positioning accuracy [33–36]. Generally, the kinematic model-based calibration involves four sequential steps: Modeling, Measurement, Identification, Correction. Modeling is a development of a mathematical model of the geometry and the robot motion. The most popular one is D-H convention [30] and other alternatives include S-model [37] and zero-reference model [38]. At the measurement step, the absolute position of the end-effector is measured from the sensors, e.g., the acoustic sensors [37], the visual sensors [34], etc. In the identification step, the parameter errors for the robot are identified by minimizing the residual position errors with different techniques [39, 40]. This final step is to implement the new model with the corrected parameters.

On the other hand, the non-kinematic calibration modeling (level 3 calibration [31]) [39, 41], which includes the dynamic factors such as the joint and the link flexibility in the calibration, increase accuracy of the robot calibration, but complicates the mathematical functions that govern the parameters relationship.

3.2.3 The image averaging techniques for denoising

The image noises are inevitably introduced in the image processing. Several image denoising techniques have been proposed so far. A good noise removal algorithm ought to remove as much noise as possible while safeguarding the edges. The Gaussian white noise has been dealt with using the spatial filters, e.g., the Gaussian filter, the Mean filter and the Wiener filter [42]. The noise reduction using the wavelet methods [43, 44] have benefits of keeping more useful details but at the expense of the computational complexity. However, depending on the selected wavelet methods, the filters that operate in the wavelet domain still filter out (or blur) some important high frequency useful information of the original image, even though more edges are preserved with the wavelet method when comparing with the spatial filter approaches.

All the aforementioned methods present ways to reduce noise in the image processing starting from a noisy image. We can approach this problem with the multiple noisy images taken from the same perspective. Assuming the same perspective ensures the same environmental conditions (illumination, temperature, etc.) that affect the image noise level. Given the same conditions, an image noise level taken at a particular time should be very similar to another image taken at a different time. This redundancy can be used for the purpose of improving image precision estimation in the presence of noise. The method that uses this redundancy to reduce noise is called signal averaging (or the image averaging in the application of the image processing) [45]. The image averaging has a natural advantage of retaining all the image details as well as reducing the unwanted noises, given that all the images for the averaging technique are taken from the same perspective. The robot's rigid end effector that holds a camera minimizes shaking and drift when shooting pictures. Furthermore, in the denoising process, the precise estimations require that the original image details to be retained. Considering these previous statements, we decided to choose image averaging over all other denoising techniques in this work.

The image averaging technique is illustrated in **Figure 3**. Assume a random, unbiased noise signal, and in addition, assume that this noise signal is completely uncorrelated with the image signal itself. As noisy images are averaged, the original true image is kept the same and the magnitude of the noise signal is compressed thus improving the signal-to-noise ratio. In **Figure 3**, we generated two random signals with the same standard deviation, and they are respectively represented by the blue and the red lines. The black line is the average of the two signals, whose magnitude is significantly decreased compared to each of the original signal. In general, we can come up with a mathematical relationship between the noise level reduction and the sample size for averaging. Assume we have $\boldsymbol{N}$ numbers of Gaussian white noise samples with the standard deviation $\boldsymbol{\sigma}$. Each sample is denoted as z_i, where $\boldsymbol{i}$ represents $\boldsymbol{i}^{th}$ sample signal. Therefore, we can acquire that:

$$var(z_i) = E\left(z_i^{\,2}\right) = \sigma \tag{17}$$

where $E(\cdot)$ is the expectation value and σ is the standard deviation of the noise signal. By averaging the N Gaussian white noise signals, we can write:

$$var\left(z_{\text{avg}}\right) = var\left(\frac{1}{N}\sum_{i=1}^{N} z_i\right) = \frac{1}{N^2}N\sigma^2 = \frac{1}{N}\sigma^2 = \left(\frac{1}{\sqrt{N}}\sigma\right)^2 \tag{18}$$

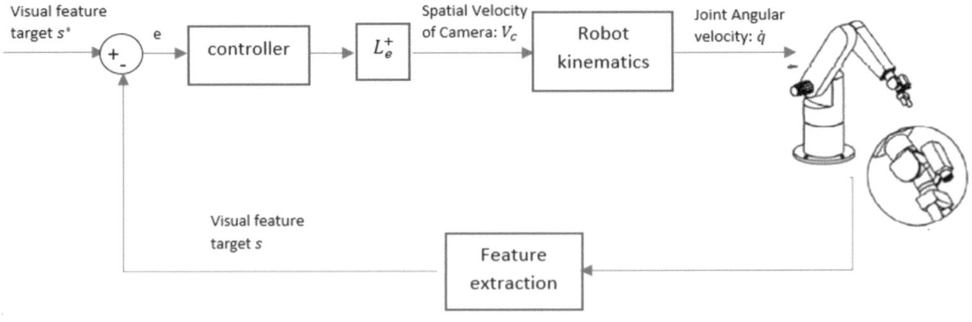

Figure 3.
An example of a noise level reduction by image averaging.

where $\mathbf{z}_{avg}$ is the average of the N noise signals. Eq. (18) demonstrates that a total number of N samples are required to reduce the signal noise level by $\sqrt{N}$. Since our goal is to reduce the image noise to within a fixed threshold (a constant number expressed as a standard deviation), a smaller variation in the original image requires much less samples to make an equivalent noise-reduction estimation. Thus, it is worthwhile for the camera to move around, rather than being stationary, in order to find the best locations where the image noise level estimation is small. In general, we can reduce the noise level as much as needed by taking more samples.

3.2.4 The dynamic errors and its modeling in the feedback control loop

The dynamic errors in a robot manipulator consist of any joint non-static errors. Among these error sources, which have the most significant effect on the robot position control accuracy, are the deviations between the actual joint rotation with its measured value from the unmodeled dynamic uncertainties, such as backlash, friction, compliance due to gears' elimination, joint or link flexibility and thermal effects. As discussed in Section 3.2.2, we can account for all these errors in the dynamic modeling step by developing a high-fidelity dynamic model, where all these parameter values could be identified through calibration. An easier way is to regard all these dynamic errors as disturbances to a manipulator control system. The control system is able to make the plant output track the desired input (the reference signal) and while simultaneously, it rejects these disturbances. The design of the robot manipulators control systems and the demonstrations of the capability of these feedback loops to reject these aforementioned dynamic errors are discussed in more detail in the later sections.

4. The topology of the multi-robotic system for accurate positioning control

In this section, we discuss the proposed control architectures for a multi-robot system, which enables the high-accuracy movement of a tool in various manufacturing scenarios by reducing the process uncertainties. Assuming at the start, all camera and robot manipulators are well calibrated by using one or multiple methods discussed in Sections 3.2.1 and 3.2.2, so that the initial camera and robot manipulators parameters are identified. Therefore, in this case, the main uncertainties include the sensor noise and the dynamic modeling errors. **Figure 4** shows the overall topology of this multi-robot system.

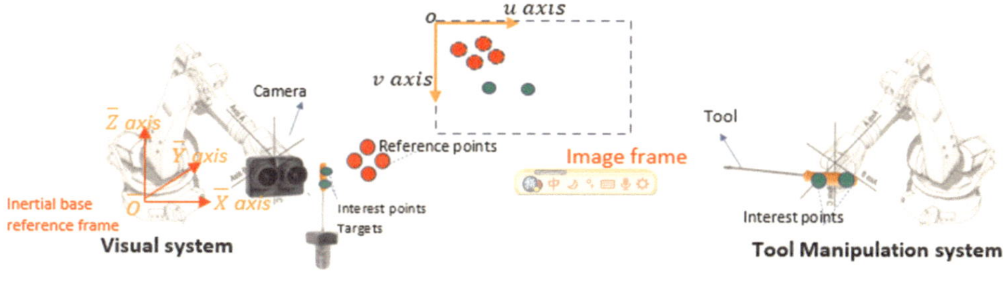

Figure 4.
The topology of the multi-robotic system for accurate positioning control.

The multi-robot system is composed of a visual system and a tool manipulation system (**Figure 4**). In the visual system, a camera is mounted on an elbow robot arm while a tool is held by the end-effector of the robot manipulator arm. The goal of the visual system is to provide precise estimation of the tool pose so that the tool manipulator can control the pose with the guidance from the visual system. Two fiducial markers (green circles or the interest points) are placed on the tool to help the computer to detect the position and the orientation of the tool. The absolute coordinates of the reference points (red circles) are known in the inertial reference frame. The reference points are placed close to the tool's target location (the target interest points) so that the reference points and the target interest points can be captured in the camera frame when the tool gets close to its target pose.

Four reference points are selected close to each other in the space. In a visual servoing problem, a location in space from which an image was taken can only be uniquely determined by at least four points, 3 points to determine a specific location and one point to determine the orientation. This is a location determination problem (LDP) using image recognition [46]. Therefore, we consider using four reference points to determine the camera pose in the 3D space. However, whenever the camera pose is fixed and known in space, the stereo camera, which can detect the depth, provides the distinct 3D location of a point from the image coordinates.

4.1 The multi-robotic system sequential control procedure

The movement control of the robot manipulators is asynchronous in the visual and the tool manipulator systems. A flowchart demonstrating this sequential control process is shown in **Figure 5**.

The first stage consists of the optimal camera pose determination and control. In this stage, the camera moves and searches for an optimal position based on the minimization of a proposed objective function, in this case, the time duration and the energy consumption, while reducing the image noise of the reference points to within an acceptable threshold. In the second stage, the camera movement adjustment control, any uncertainties occur in the movement of camera from the last stage is eliminated by an eye-in-hand visual servoing controller. After the movement adjustment, the camera is kept static and provides precise estimations of the tool position. In the last stage, the high-accuracy tool manipulation control, the tool movement is controlled and guided to the target pose location by an eye-to-hand visual servoing controller. Each control method and architecture are discussed in the sections below.

4.2 The optimal camera pose determination process and its control architecture

Figure 6 shows how the optimal pose of the camera is determined from a single picture taken at different perspectives. The uncertainty in the image processing is spatially related. As the camera moves in space, the combined factors (the light conditions, the temperature, etc.) that affect the image processing changes and with

Figure 5.
The flow chart of the sequential control procedure.

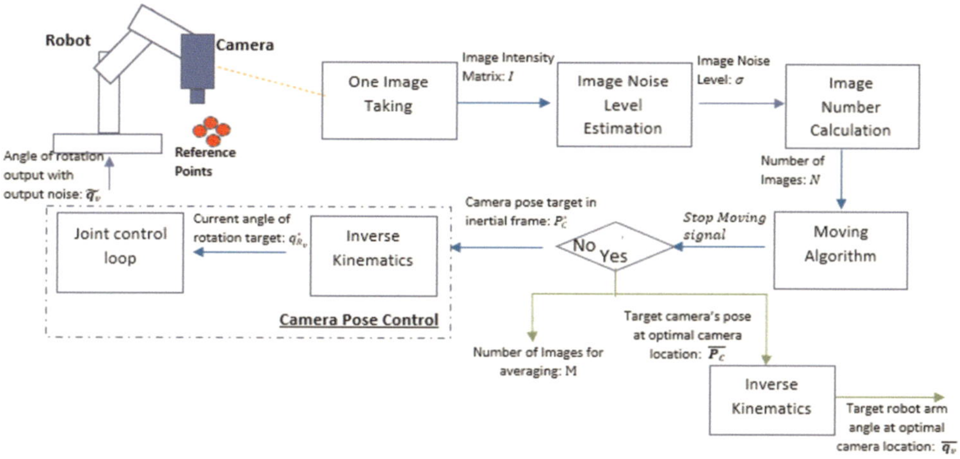

Figure 6.
The optimal camera pose determination process and its control architecture.

these changes, the uncertainty level in the estimation changes accordingly. In this work, we propose to apply image averaging [45] to reduce the uncertainty level in the pose estimation. As discussed in Section 3.2.3, the number of images required for the averaging increases by a factor of 2 for reducing the uncertainty level by the square root of the same factor. In order to reduce the energy consumption and the time duration in this photo taking process, it is necessary to first determine the location where the image averaging should take place before the camera actually starts to take multiple photos.

In **Figure 6**, in the first stage, the camera takes a single picture. In the second stage, we compute the image intensity matrix $\boldsymbol{I}$ from that photo and then, we estimate the noise level σ across the image by a previously developed algorithm, see [47]. In the third stage, we calculate the uncertainty level from the image noise level and generates the number of images $\boldsymbol{N}$ required to reduce this uncertainty within a prescribed threshold. In the fourth stage, utilizing a moving algorithm, which is designed as a part of this work, the current camera target pose, $\boldsymbol{P_C}$ is commanded. In the fifth stage, the camera pose controller guides the camera to the target pose location using the encoders that measure joints rotational angles, $\tilde{q}_v$. These five stages are repeated until the movement algorithm instructs the camera to stay in the current pose. Then, this current target pose is the optimal pose $\overline{\boldsymbol{P_C}}$ of the camera where the total energy consumption and the time duration is minimized. The output $\overline{\boldsymbol{q_v}}$ is the target joint angles of visual manipulator system at the optimal pose of the camera. If for any reason, such as uncompensated uncertainties, the current pose is not the same as the optimal pose, $\overline{\boldsymbol{P_C}}$, then the camera movement adjustment control, presented in the next section, will reduce this error. In addition, $\boldsymbol{M}$ is the number of pictures needed for the averaging at the optimal pose location of the camera.

4.3 The camera movement adjustment control block diagram

We propose a control method with its associated block diagram for the camera movement adjustment as shown in **Figure 7**. The role of this feedback control is to deal with the errors occurred in the dynamics and the measurements of the previous stage.

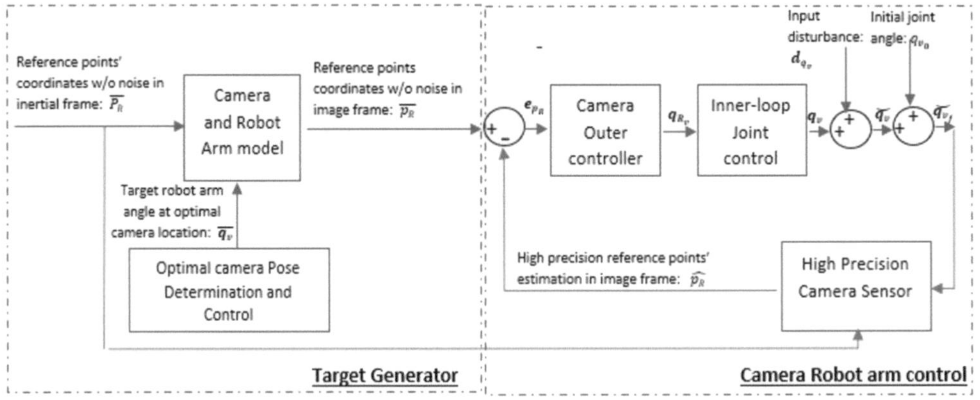

Figure 7.
The camera movement adjustment control block diagram.

In **Figure 4**, four reference points whose absolute positions are known in the space, are selected close to the tool target pose. The fiducial markers are placed on the reference points, so that their location can be recognized and estimated in a 2D image coordinate frame using computer vision. From the kinematic model of the robot arm and the camera, the image coordinates of the reference points can be calculated online, and those coordinates are used as the targets for a cascaded control loop and are noted as $\overline{p_R}$ in **Figure 7**. After applying the image averaging technique (Section 3.2.3), we can obtain a precise estimation of the current position of the reference points and are noted as $\widehat{p_R}$ in **Figure 7** in image coordinates from the computer vision. Therefore, any deviation between $\overline{p_R}$ and $\widehat{p_R}$ could be the result of some uncertainties, such as the joint compliances, which are not compensated by the joint control loop as shown in **Figure 6**. The cascaded controller is similar to the image-based visual servoing scheme (IBVS), as discussed in Section 3.1. The inner-loop control strategy in this part is also very similar to the joint control in the camera pose control in **Figure 6**, and its control design and simulation results are discussed in Section 5.

4.4 The high-accuracy tool manipulation control block diagram

We propose a control strategy with its associated control block diagram for the tool manipulation system as shown in **Figure 8**. The control algorithm in this block diagram is a combination of a feedforward and a feedback control.

The feedforward control loop is an open loop which brings the tool as close to the target position as possible in the presence of the input disturbance, d_{q_m}. In the inner joint control loop, the noise sources may originate from the low fidelity cheap encoder joint sensors and the dynamic errors from the joint, e.g., compliances. All sources of noise from the joint control loop are combined and modeled as an input disturbance, d_{q_m}, to the outer control loop. The outputs of the feedforward are the reference joint angles of rotations, $q_{R_{m_{feedforward}}}$, which are added to the outer feedback controller outputs, $q_{R_{m_{feedback}}}$, and set as the targets for the joint control inner-loop. The function of the forward kinematics is to transform a set of current joint angles of the tool manipulator to the current pose of the tool on the end-effector using a kinematic model of the robot arm.

Movement of the tool can be adjusted with high accuracy by the feedback control loop. The feedback control loop rejects the input disturbance, d_{q_m}, and minimizes the

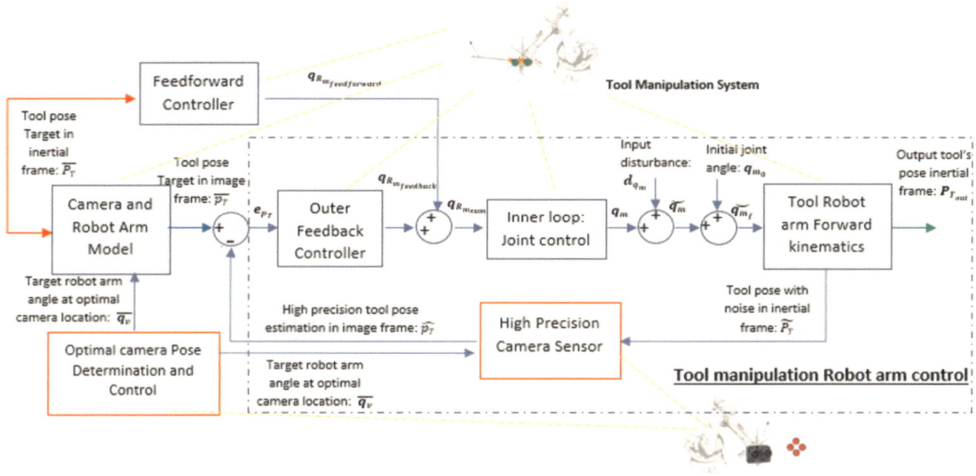

Figure 8.
The high-accuracy tool manipulation control block diagram.

error between the tool pose target, $\overline{p_T}$, in the image frame and the high precision estimation tool pose from the camera sensor, $\widehat{p_T}$. The pose in a robot system modeled in Cartesian inertial base frame consists of six degrees of freedom, i.e., three translations and three rotations. Therefore, in order to have a full control of the tool pose, the camera in the feedback control loop requires to measure the image coordinates of at least two interests points on the tool.

The feedforward and feedback controllers work simultaneously to move the tool to the target pose location in the tool manipulation system. The combined target $q_{R_{msum}}$ are the inputs to the joint control loop so that both controllers manipulate the tool pose. The benefit of designing both feedback and feedforward controls for the manipulation system is to reduce the time duration. If only feedback control is utilized, the pose estimation generated from the visual system requires taking multiple pictures and makes the tool movement very slow. We can divide the task of the tool movement control into two stages. In the first stage, under the action of the feedforward control, the tool moves to an approximate location that is close to the desired destination. In the second stage, the feedback controller moves the tool to the precise target location using the tool pose estimation from the camera. In addition, the camera has a range of view and can only detect the tool and measure its 2D feature as $\widehat{p_T}$ when it is not far away from the target. When the tool is moving from a location that is not in the camera range of view, we must estimate the feature as $\widetilde{p_T}$ until the tool moves into the range of view (this point will be discussed in detail in Section 7). It should be noted that only the feedback controller has the ability to compensate for uncertainties.

This control topology is an analogy to the macro-micro manipulation in the current industry trends where the large-scale robots are used for the approximate positioning, while the small-scale robots are utilized for the precise positioning [2].

4.5 The high precision camera sensor model

As shown in **Figures** 7 and **8**, the high precision camera sensor model provides high precision estimations in the feedback loop of the tool manipulation system control and the camera movement adjustment control. The camera robot arm model, which is shown in both **Figures 7** and **8**, is the target generator that transforms the target in the inertial frame to the target locations in the image frame. The mathematical model of the camera, which is utilized in the visual robot arm and in the feedback loop to generate the required position estimation, has an equivalent Hardware-In-the-Loop (HIL) model as shown in **Figure 9**.

Figure 9 shows the details of high precision camera model and its equivalent HIL model. The upper configuration is the mathematical model that is used in the simulation to generate image coordinates and to design the outer-loop controller in the robot arm control loop. However, in real application, the lower HIL configuration replaces this mathematical model. In the HIL model, the image processing will make an estimation of the tool pose with high precision.

5. The inner-loop control design

In this part, we will present the control system design of the camera pose control and the inner joint control loops for both the camera movement adjustment and the tool manipulation, which have been introduced in Section 4. A simulation scenario is also presented in this section.

We express an equation for the 6 DOF manipulator including the dynamics of the robot manipulator and the actuators (DC motor) in the following simplified form [28]:

$$(D(q) + J)\ddot{q} + \left(C(q, \dot{q}) + \frac{B}{r}\right)\dot{q} + g(q) = u \tag{19}$$

where $\boldsymbol{D(q)}$ and $\boldsymbol{C(q, \dot{q})}$ are $\boldsymbol{6 \times 6}$ inertial and Coriolis matrices respectively. $\boldsymbol{J}$ is a diagonal matrix expressing the sum of actuator and gear inertias. $\boldsymbol{B}$ is the damping factor, and $\boldsymbol{r}$ is the gear ratio, $\mathbf{g}(\boldsymbol{q})$ is the term for potential energy, u is the 6×1 input

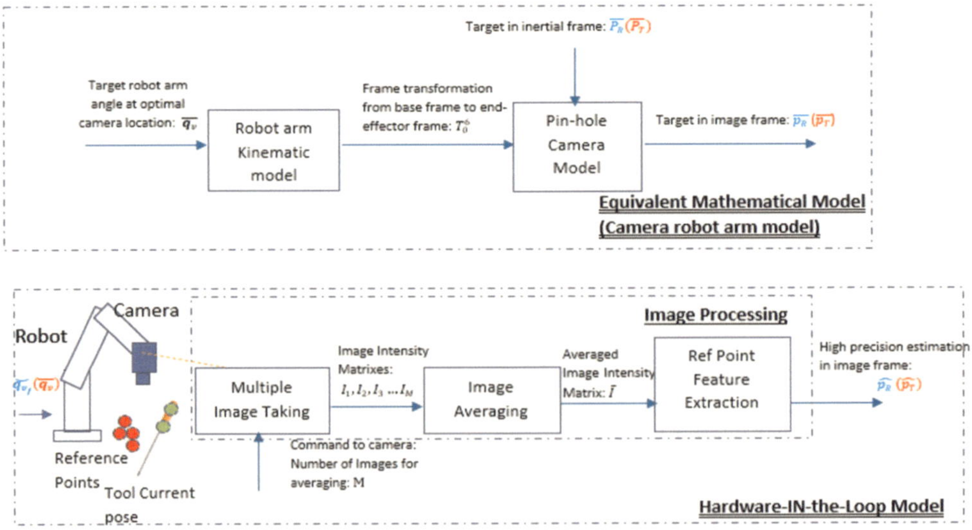

Figure 9.
*The high precision camera sensor. Note: Blue are signals used in the control system of **Figure 7** and red are signals used in the control system of **Figure 8**.*

vector, and q is the 6×1 generalized coordinates (in this paper, q is the 6×1 joint angle vector).

We can simplify Eq. (19) as follows:

$$M(q)\ddot{q} + h(\mathrm{q}, \dot{\mathrm{q}}) = \mathrm{u} \tag{20}$$

with

$$M(q) = D(q) + J \tag{21}$$

$$h(\mathrm{q}, \dot{\mathrm{q}}) = \left(C(q, \dot{q}) + \frac{B}{r}\right)\dot{q} + \mathrm{g}(q) \tag{22}$$

If we transform the control input as following:

$$\mathrm{u} = M(q)v + h(\mathrm{q}, \dot{\mathrm{q}}) \tag{23}$$

where v is a virtual input. Then we substitute for u in Eq. (20) using (23), and since $M(q)$ is invertible, we will have a reduced system equation as follows:

$$\ddot{q} = v \tag{24}$$

This transformation is so-called feedback linearization technique with the new system equation given in Eq. (24). This equation represents 6 uncoupled double integrators. The overall feedback linearization method is illustrated in **Figure 10**. In this control block diagram, we force the joint angle q to follow the target joint angle q_R so that the output pose, P, can follow the target pose, $\overline{P}$. P, $\overline{P}$, q, and q_R are all vectors with six elements (each element corresponds to a joint position or angle). The Nonlinear interface transform the linear virtual control input v to the nonlinear control input u by using Eq. (23). The output of the manipulator dynamic model, the joint angles, q, and their first derivatives, $\dot{q}$, are utilized to calculate $M(q)$ and $h(\mathrm{q}, \dot{\mathrm{q}})$ in the Nonlinear interface. The linear joint controller is designed using Youla parameterization technique [9] to control the nominally linear system in Eq. (24).

The design of a linear Youla controller with nominally linear plant is presented next.

Since the transfer functions between all inputs to outputs in (24) are the same and decoupled, we can first design a SISO (single input and single output) controller and use the multiple of the same controller for a six-dimension to obtain the MIMO (Multiple Input Multiple Output) version. In other words, we first design a controller $\mathbf{G}_{cSISO}$ that satisfies:

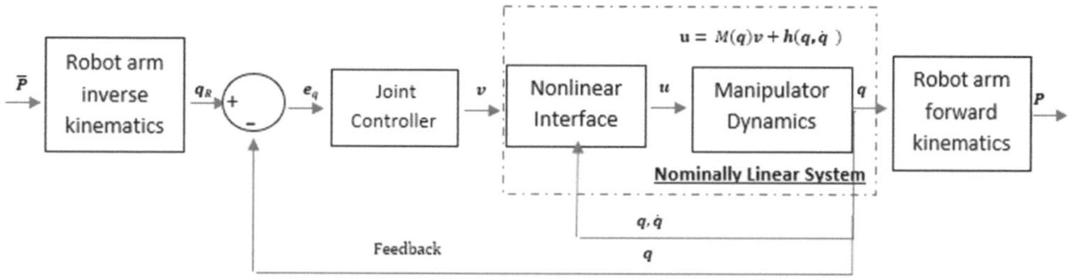

Figure 10.
The block diagram of feedback linearization Youla control design used for the joint control loop.

$$v_{SISO} = \ddot{q}_{SISO} \tag{25}$$

where v_{SISO} is a single input to a nominally linear system and $\ddot{q}_{SISO}$ is the second order derivative of a joint angle. The controller system in **Figure 10** can be then written as:

$$G_{c_{sys}} = G_{cSISO} \cdot I \tag{26}$$

where I is a 6×6 identity matrix. We can design the SISO controller G_c using Youla parameterization technique [48]. The transfer function of the SISO nominally linear system from (24) is:

$$G_{p_{SISO}} = \frac{1}{s^2} \tag{27}$$

Note that $G_{p_{SISO}}$ has two BIBO (Bounded Input Bounded Output) unstable poles at origin. To ensure internal stability of the feedback loop, the closed loop transfer function, T_{SISO}, should meet the interpolation conditions [39]:

$$T_{SISO}(s = 0) = 1 \tag{28}$$

$$\left.\frac{dT_{SISO}}{ds}\right|_{s=0} = 0 \tag{29}$$

We compute a Youla transfer function: Y_{SISO}, using the following relationship,

$$T_{SISO} = Y_{SISO} G_{p_{SISO}} \tag{30}$$

The T_{SISO} is designed so that it satisfies the conditions in (28) and (29). The sensitivity transfer function, S_{SISO}, is then calculated as follows:

$$S_{SISO} = 1 - T_{SISO} \tag{31}$$

Without providing the design details, we refer the interested reader to [38], the closed-loop transfer function should be in the following form to satisfy the interpolation conditions:

$$T_{SISO} = \frac{(3\tau s + 1)}{(\tau s + 1)^3} \tag{32}$$

Where τ specifies the pole and zero locations and represents the bandwidth of the control system. We can tune τ so that the response can be fast with less-overshoot.

Then we can derive G_{cSISO} from relationships between the closed-loop transfer function, T_{SISO}, the sensitivity transfer function, S_{SISO}, and the Youla transfer function, Y_{SISO}, in Eqs. (30)–(32):

$$Y_{SISO} = T_{SISO} G_{p_{SISO}}^{-1} = \frac{s^2(3\tau^2 s + 1)}{(\tau s + 1)^3} \tag{33}$$

$$S_{SISO} = 1 - T_{SISO} = \frac{s^2(\tau^3 s + 3\tau^2)}{(\tau s + 1)^3} \tag{34}$$

$$G_{cSISO} = Y_{SISO}S_{SISO}^{-1} = \frac{3\tau^2 s + 1}{\tau^3 s + 3\tau^2} \tag{35}$$

From Eq. (35), we can compute a MIMO controller as follows:

$$G_{c_{sys}} = \frac{3\tau^2 s + 1}{\tau^3 s + 3\tau^2} \cdot I_{6\times6} \tag{36}$$

Eq. (36) provides the desired controller, which is used as the joint controller, as shown in **Figure 10**. This configuration is precisely the inner joint control loop in both the visual and the manipulator systems as shown in **Figures** 7 and 8.

Figure 11 shows the simulation results for the case with no disturbance. The target position and the orientation of the end-effector are selected to be $\begin{bmatrix} \overline{X} \\ \overline{Y} \\ \overline{Z} \end{bmatrix} = \begin{bmatrix} 1.7157m \\ 1.0191m \\ 0.7518m \end{bmatrix}$ and $\begin{bmatrix} \overline{n} \\ \overline{s} \\ \overline{a} \end{bmatrix} = \begin{bmatrix} -0.425 & 0.87 & 0.25 \\ 0.8361 & 0.2714 & 0.4767 \\ 0.3469 & 0.4116 & -0.8428 \end{bmatrix}$, where $\left[\overline{X}, \overline{Y}, \overline{Z}\right]^T$ is the absolute position coordinate of the center of the end effector in the inertial frame and $\overline{n}, \overline{s}, \overline{a}$ represent respectively the end-effector's directional unit vector of the yaw, pitch and roll in the inertial frame. Therefore, the corresponding target angles of rotations are $q_R = [30°, 60°, -45°, 15°, 45°, 90°]$. For this simulation, we have designed the control system with the bandwidth of 100 rad/s. In the following three plots, solid lines represent the responses for the end-effector position of each joint and the end-effector orientation respectively, and the dashed lines are the targets. Specifically, the orientation response of the end-effector is the vector that tangent to the curve in the second plot at each point in **Figure 11**. The simulation results show that all responses of the controlled system will be able to reach their final/steady state values within 0.1 second with no (or little) overshoots.

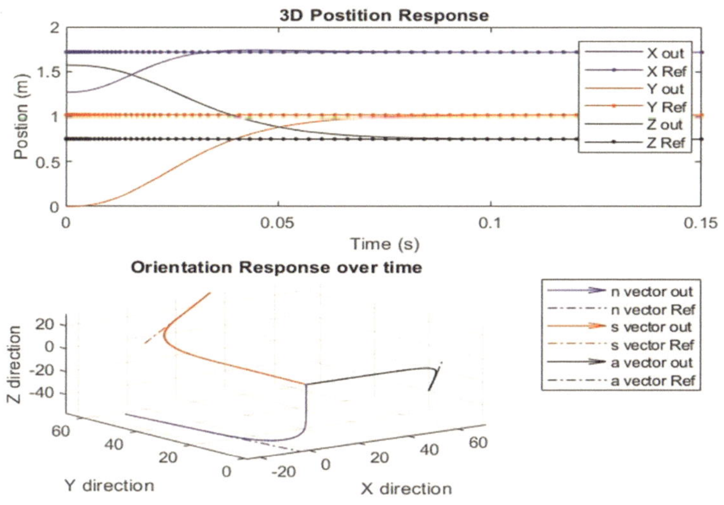

Figure 11.
The simulation results for the end-effector response to an arbitrary selected trajectory.

6. The cascaded SISO outer-loop design for the camera movement adjustment control system

As introduced in Section 4.1, we use the Image Based Visual Stereo (IBVS) as the framework for cascaded control design of both the visual and the manipulation system. The inner-loop is the joint control loop, as discussed previously. In the IBVS, an outer feedback control is designed in addition to the inner feedback control so that the 2D visual features can be compared and matched. Therefore, a camera model is required in the feedback loop to map the joint angles to the object visual features (e.g., 2D coordinates of the object in the image frame). To start from a simple case, we discuss the requirements for the design of the cascaded control for a SISO system. Assume, we only measure and try to control one feature: the coordinate of the object in one axis, then the overall cascaded control diagram of this configuration is shown in **Figure 12**.

$\overline{u_R}$ is the target of u coordinate of a point (one reference point in visual system). $G_{C_{out}}$ is the outer-loop controller, which provides the target joint angle q_{R_v} to the robot arm based on the image coordinate error e_{u_R}. A combined uncertainty signal, d_{q_v} (e.g., sensor noise, backlash, friction, and compliance due to gear reduction in the joint) is added to the joint angle of rotation q_v, the output of the inner control loop. Furthermore, the initial joint angle of robot manipulator, q_{v0}, is added to the output joint angle with disturbance, $\overset{\smile}{q_v}$, to generate the final joint angle $\overset{\smile}{q_{v_f}}$.

6.1 The camera sensor model for the SISO control design

In this section, we are going to present a simplified mathematical model of the camera. Let us consider a one-link robot manipulator with a camera that can be rotated around the Z-axis, as shown in **Figure 13**. Let us also consider a point that is located in the $X-Y$ plane.

We use a Cartesian coordinate frame $OXYZ$, which is attached at the camera center. The X-axis is perpendicular to the camera lens and the coordinate frame rotates around Z-axis. A point whose coordinate is $(X_P, Y_P, 0)$ in the rotational frame is projected on the image $u-v$ plane with its coordinates as $(u, 0)$. In addition, an inertial frame $\overline{OXYZ}$ is shown at the base of the manipulator in **Figure 13**.

We derive the equations for a pin-hole camera model (a monocular case in stereo model in Section 2.1). Assuming the skew coefficient and image coordinate offsets are zero ($\boldsymbol{s_c = u_0 = v_0 = 0}$), then from Eq. (1), the one-dimensional image coordinate u can be written as:

$$u = \frac{f_u Y_P}{X_P} \tag{37}$$

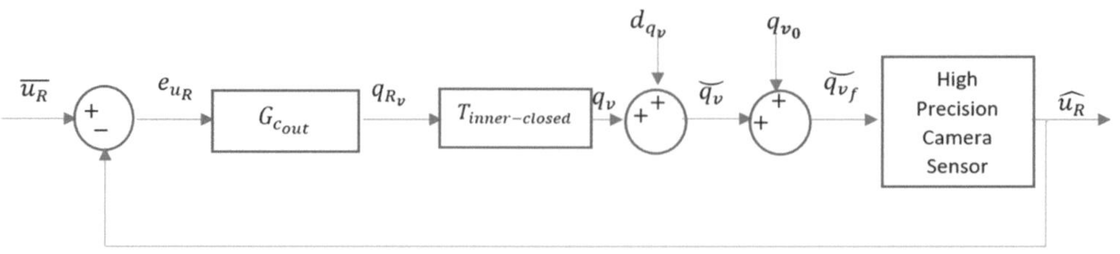

Figure 12.
Control block diagram for the SISO camera movement adjustment system.

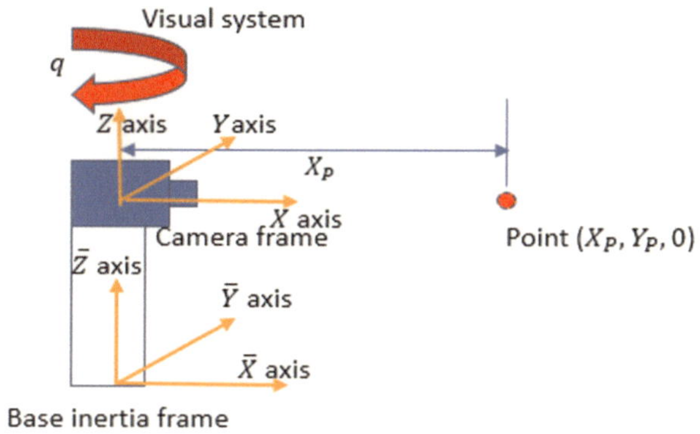

Figure 13.
One-link manipulator with a camera.

The coordinate system, located at the camera center, rotates with the camera. Assume the camera starts at the position where the X-axis is parallel to the $\overline{X}$-axis of the inertial coordinate system and positive angle is defined when the camera rotates in a clockwise direction with an angle q around Z-axis. After the frame rotates q clockwise, the new coordinates of the point in the new coordinate frame $(X'_P, Y'_P, 0)$ can be computed as:

$$\begin{bmatrix} X'_P \\ Y'_P \end{bmatrix} = \begin{bmatrix} \cos(q) & -\sin(q) \\ \sin(q) & \cos(q) \end{bmatrix} \begin{bmatrix} X_P \\ Y_P \end{bmatrix} \tag{38}$$

Combining Eqs. (37) and (38), the new image coordinate, u', can be calculated as:

$$u' = \frac{f_u Y'_P}{X'_P} = f_u \frac{X_P sin(q) + Y_P cos(q)}{X_P cos(q) - Y_P sin(q)} \tag{39}$$

Let us write the coordinates (X_P, Y_P) in the polar coordinates, such that,

$$R = \sqrt{X_P{}^2 + Y_P{}^2}$$

and

$$X_P = Rcos(\varphi), Y_P = Rsin(\varphi),$$

where

$$\varphi = tan^{-1}(Y_P/X_P) \tag{40}$$

Therefore, Eq. (39) can be written in the polar form as:

$$u' = f_u \frac{Rsin(\varphi + q)}{Rcos(\varphi + q)} = f_u tan(\varphi + q) \tag{41}$$

The angle φ is the angle of the point with respect to the $\overline{X}$-axis and the angle q is is already defined. Eq. (41) mathematically expresses a model of the camera sensor shown in the block diagram of **Figure 13**.

6.2 Outer-loop controller design

In the previous sections, we presented the results for designing inner joint loop controllers using Youla parameterization method. In the following two sections, we will discuss the design of the outer-loop controllers by using two different approaches:

a. Feedback linearization

b. Model linearization

6.2.1 Feedback linearization

Let us relook at the cascaded block diagram in **Figure 12**. The inner closed loop transfer function is already derived in (32):

$$T_{inner-closed} = \frac{q_v(s)}{q_{R_v}(s)} = \frac{(3\tau_{in}s + 1)}{(\tau_{in}s + 1)^3} \tag{42}$$

Notice we replaced τ with τ_{in} to indicate bandwidth of inner-closed loop. Using Eq. (41) and considering the block diagram in **Figure 12**, we can write:

$$\widehat{u_R} = f_u tan\left(\varphi + \smile q_{vf}\right), with\ \varphi = tan^{-1}(Y_P/X_P) \tag{43}$$

We can rewrite (42) and (43) in time domain, by introducing an intermediate variable or state, W, as:

$$\tau_{in}\dddot{W} + 3\tau_{in}{}^2\ddot{W} + 3\tau_{in}\dot{W} + W = q_{R_v} \tag{44}$$

$$f_u tan\left(\varphi + q_{v0} + 3\tau_{in}\dot{W} + W\right) = \widehat{u_R} \tag{45}$$

Eqs. (44) and (45) describe a nonlinear third order system, where W is the state, q_{R_v} is the input and $\widehat{u_R}$ is the output. We can use feedback linearization method to design the outer-loop controller by taking the second order derivative of (45) and combine with (44) to obtain:

$$\ddot{\widehat{u_R}} = R(W, \dot{W}, \ddot{W}) + G(W, \dot{W}, \ddot{W})q_{R_v} \tag{46}$$

where

$$R(W, \dot{W}, \ddot{W}) = 2f_u cos^{-2}(\varphi + q_{v0} + 3\tau_{in}\dot{W} + W)\tan(\varphi + q_{v0} + 3\tau_{in}\dot{W} + W)$$
$$\left(3\tau_{in}\ddot{W} + \dot{W}\right)^2 - f_u cos^{-2}(\varphi + q_{v0} + 3\tau_{in}\dot{W} + W)\left(8\ddot{W} + \frac{9}{\tau_{in}}\dot{W} + \frac{3}{\tau_{in}{}^2}W\right) \tag{47}$$

$$G(W, \dot{W}, \ddot{W}) = f_u cos^{-2}(\varphi + q_{v0} + 3\tau_{in}\dot{W} + W)\frac{3}{\tau_{in}{}^2} \tag{48}$$

We can transform or map these variables so that the nonlinear system in (47)–(48) can be written as an equivalent linear state-space representation as follows:

$$\varepsilon_1 = f_u tan\left(\varphi + q_{v0} + 3\tau_{in}\dot{W} + W\right) \tag{49}$$

$$\varepsilon_2 = \dot{\varepsilon_1} = f_u cos^{-2}\left(\varphi + q_{v0} + 3\tau_{in}\dot{W} + W\right) \cdot \left(3\tau_{in}\ddot{W} + \dot{W}\right) \tag{50}$$

We can write the state-space form of (49) and (50) as:

$$\dot{\varepsilon} = \begin{bmatrix} \dot{\varepsilon_1} \\ \dot{\varepsilon_2} \end{bmatrix} = \begin{bmatrix} 0 & 1 \\ 0 & 0 \end{bmatrix} \begin{bmatrix} \varepsilon_1 \\ \varepsilon_2 \end{bmatrix} + \begin{bmatrix} 0 \\ 1 \end{bmatrix} U \tag{51}$$

$$\widehat{u_R} = [1\ 0] \begin{bmatrix} \varepsilon_1 \\ \varepsilon_2 \end{bmatrix} \tag{52}$$

where

$$U = G(W, \dot{W}, \ddot{W}) q_{R_v} + R(W, \dot{W}, \ddot{W}) \tag{53}$$

Transform the state-space representation back to the transfer function form, we can write:

$$Gp_{nominal} = \frac{\widehat{u_R}(s)}{U(s)} = C(sI - A)^{-1}B = \frac{1}{s^2} \tag{54}$$

where

$$A = \begin{bmatrix} 0 & 1 \\ 0 & 0 \end{bmatrix}, B = \begin{bmatrix} 0 \\ 1 \end{bmatrix}, and\ C = [1\ 0] \tag{55}$$

Since the $Gp_{nominal}$ is the same as the plant transfer function in (27), the design of Youla controller for this linear system is similar to (28)–(35). Therefore, the transfer function of the outer-loop controller can be written as:

$$G_{C_{out}} = \frac{3\tau_{out}{}^2 s + 1}{\tau_{out}{}^3 s + 3\tau_{out}{}^2} \tag{56}$$

τ_{out} determines the pole and zero locations of closed-loop transfer function of the outer-loop controller and therefore, represents the bandwidth of outer-loop controller. We must make sure that $\tau_{out} > \tau_{in}$ so that inner-loop responds faster than the outer-loop in the cascaded control design strategy. The overall block diagram is shown in **Figure 14**.

The states $W, \dot{W}, \ddot{W}$ are computed from the rotational angle and its derivative $\breve{q}_{v_f}$ and $\dot{\breve{q}}_{v_f}$. From Eq. (42), we can obtain the following relationship:

$$\breve{q}_{v_f} = 3\tau_{in}\dot{W} + W \tag{57}$$

Therefore, the transfer function from W to $\breve{q}_{v_f}$ can be written as:

$$W(s) = \frac{1}{3\tau_{in}s + 1} \breve{q}_{v_f}(s) \tag{58}$$

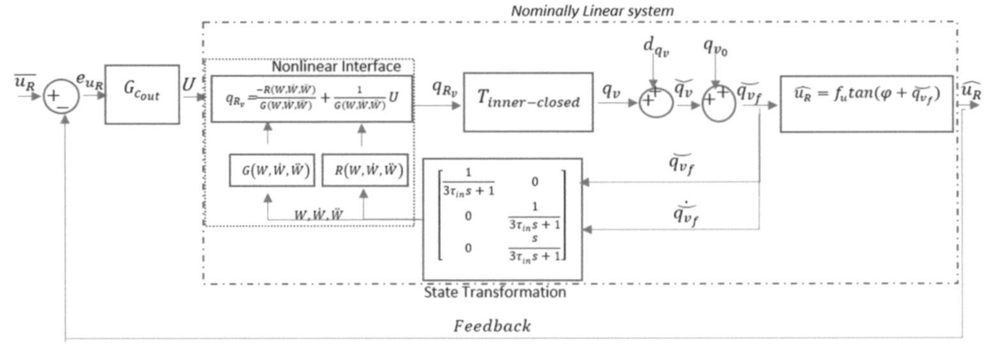

Figure 14.
The block diagram of the SISO outer-loop control with feedback linearization for robot arm movement adjustment.

Also, we can obtain the transfer function of $\dot{W}(s)$ and $\ddot{W}(s)$ as:

$$\dot{W}(s) = \frac{1}{3\tau_{in}s+1}\dot{\breve{q}}_{v_f}(s) \tag{59}$$

$$\ddot{W}(s) = \frac{s}{3\tau_{in}s+1}\dot{\breve{q}}_{v_f}(s) \tag{60}$$

Eqs. (58)–(60) are shown as the state transformation block in **Figure 14**.
It is worthwhile to note that the equation in the nonlinear interface:

$$q_{R_v} = \frac{-R(W,\dot{W},\ddot{W})}{G(W,\dot{W},\ddot{W})} + \frac{1}{G(W,\dot{W},\ddot{W})}U \tag{61}$$

is only defined when $G(W,\dot{W},\ddot{W}) = f_u cos^{-2}\left(\varphi + q_{v_0} + q_v\right)\frac{3}{\tau_{in}{}^2} \neq 0$. It can be shown that this is always not equal to zero and $\varphi + q_{v_0} + q_v \neq \pm\frac{\pi}{2}$. This is always true because any angle of rotation should be within half of camera's angle of view α; that is $\left|\varphi + \breve{q}_{v_f}\right| \leq \frac{\alpha}{2} < \frac{\pi}{2}$. In addition, α is always less than π for any camera type. Therefore, the controller works for the entire range independent of the camera type.

6.3 Model linearization

We can deal with the nonlinear system by linearizing the system first and then design a linear controller using the system transfer function. In **Figure 12**, $T_{inner-closed}$ transfer function is given in (42) and the nonlinear form of the camera model is provided in (43). The overall dynamic system combines the inner-loop and the camera model, which will be linearized so that the combined dynamic system will then be linear. Next, we linearize, the camera model, (43), around an equilibrium point $\breve{q}_{v_0}$:

$$\widehat{u_R} = f_u cos^{-2}\left(\varphi + \breve{q}_{v_0}\right)\left(\breve{q}_{v_f} - \breve{q}_{v_0}\right) + f_u \tan\left(\varphi + \breve{q}_{v_0}\right) \tag{62}$$

If we assume $\breve{q}_{v_0} = 0$, then:

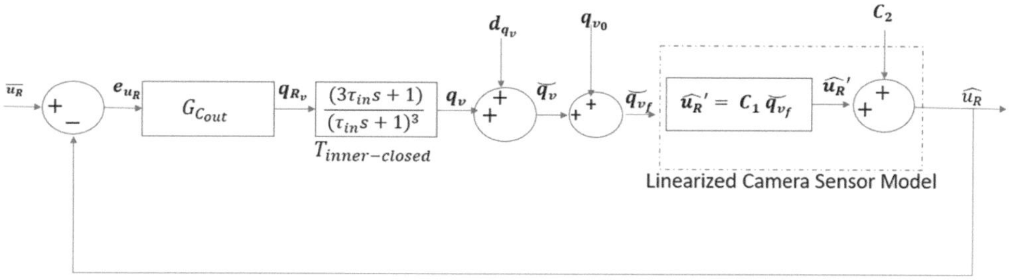

Figure 15.
The block diagram of the SISO control with the linearized camera model for robot arm movement adjustment.

$$\widehat{u_R} = f_u \cos^{-2}(\varphi)\, \breve{q_{v_f}} + f_u \tan(\varphi) \tag{63}$$

Assuming $C_1 = f_u \cos^{-2}(\varphi)$, $C_2 = f_u \tan(\varphi)$, therefore, Eq. (63) can be rewritten as:

$$\widehat{u_R} = C_1\, \breve{q_{v_f}} + C_2 \tag{64}$$

Let us define $\widehat{u_R}' = \widehat{u_R} - C_2$, then, the overall block diagram of the linearized system is shown in **Figure 15**.

The plant transfer function is derived as:

$$G_{p_{linear}} = \frac{\widehat{u_R}'}{q_{R_v}} = C_1 \frac{(3\tau_{in}s + 1)}{(\tau_{in}s + 1)^3} \tag{65}$$

The design of a Youla controller is trivial in this case as all poles/zeros of the plant transfer function in (65) are located in the left half-plane, and therefore, they are stable. In this case, we can shape the closed loop transfer function, T_{out}, by selecting a Youla transfer function: Y_{out} so that the closed loop transfer function, T_{out}, will not contain any plant poles and zeros. All poles and zeros in the original plant can be canceled out and new poles and zeros can be added to shape the closed-loop system. Let us select a Youla transfer function so that the closed-loop system behaves like a second order Butterworth filter, such that:

$$Y_{out} = \frac{1}{G_{p_{linear}}} \frac{\omega_n^2}{(s^2 + 2\zeta\omega_n s + \omega_n^2)} \tag{66}$$

then:

$$T_{out} = \frac{\omega_n^2}{(s^2 + 2\zeta\omega_n s + \omega_n^2)} \tag{67}$$

where ω_n is called natural frequency and approximately sets the bandwidth of the closed–loop system. We need to make sure the bandwidth of the outer-loop is smaller than the inner-loop, i.e., $1/\omega_n > \tau_{in}$. ζ is called the damping ratio, which is another tuning parameter.

Using Eqs. (34) and (35), we can calculate the sensitivity transfer function, S_{out}, and the controller transfer function, $G_{c_{out}}$, of the outer-loop in cascaded control design as:

$$S_{out} = 1 - T_{out} = \frac{s^2 + 2\zeta\omega_n s}{(s^2 + 2\zeta\omega_n s + \omega_n{}^2)} \tag{68}$$

$$G_{C_{out}} = Y_{out} S_{out}{}^{-1} = \frac{1}{C_1} \frac{(\tau_{in}s + 1)^3}{(3\tau_{in}s + 1)} \frac{\omega_n{}^2}{(s^2 + 2\zeta\omega_n s)} \tag{69}$$

6.4 The simulation of the cascaded SISO closed-loop system for camera movement adjustment

In this section, we are going to compare the closed-loop response results of the cascaded control system where the outer-loop controllers are designed using the two aforementioned methods: feedback linearization (Section 6.2) and model linearization (Section 6.3). The simulation results are obtained with the original nonlinear camera model (43). In addition, for the linearized plant approach, we will also illustrate how varying the damping ratio ζ affects the responses. For both methods, we chose the bandwidth of the inner-loop as $100 rad/s$ and the bandwidth of the outer-loop as $10 rad/s$.

We compare the simulation responses by choosing six different damping ratios. Two are chosen as the overdamped systems ($\zeta > 1$), one is chosen as a critically damped system ($\zeta = 1$), and three as the underdamped systems ($\zeta < 1$). We have simulated four cases and compared all six systems for each case. Each case is different due to varying the initial angle φ (see Eq. (40)) and the input disturbance d_{q_v}. Two cases are simulated without the input disturbance while the other two are simulated with the disturbance to compare the robustness of the controlled system.

The step responses of the image coordinate $\widehat{u_R}$ is shown in **Figures 16–19**. The intrinsic camera parameters are selected to be: $f_u = 2.8\ mm$ (Focal length) and $\alpha = 120°$ (Angle of view). In cases 1 and 2, the responses of feedback linearization are displayed in black dashed lines while all other lines are the responses of the controller that is designed with the linearized plant and varying the damping ratio ζ.

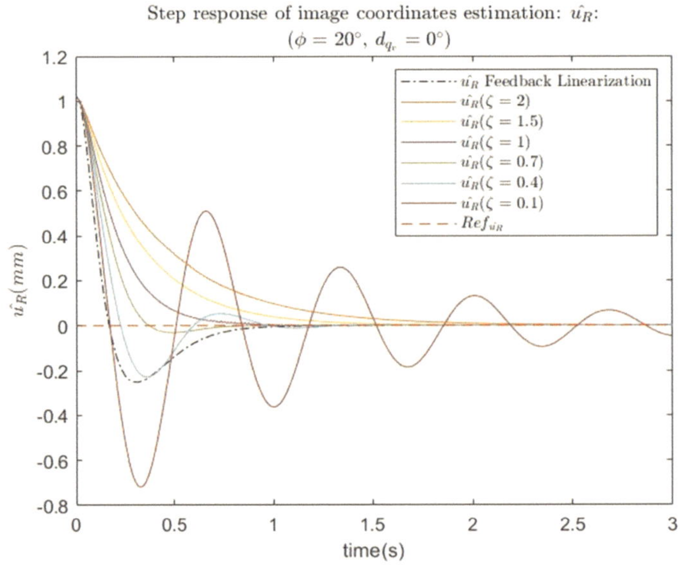

Figure 16.
Step responses of $\hat{u_R}$ *for the case 1:* $\varphi < \frac{\alpha}{2} = 20°, d_{q_v} = 0°$.

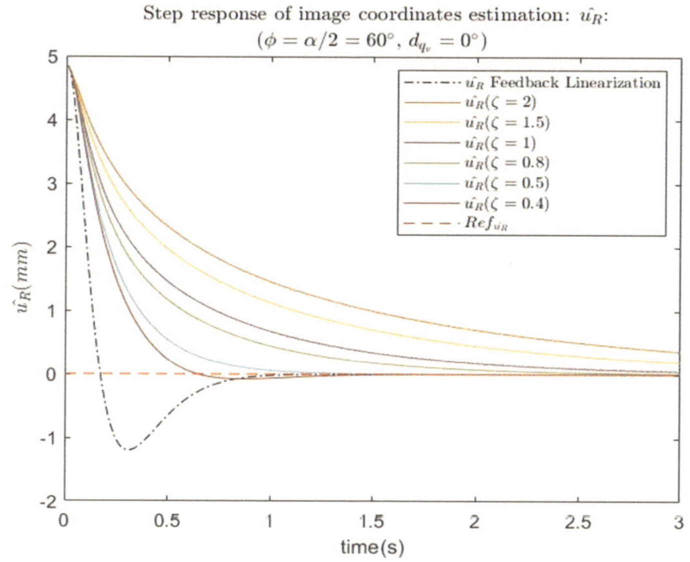

Figure 17.
Step responses of $\hat{u_R}$ *for the case 2:* $\varphi = \frac{\alpha}{2} = 60°, d_{q_v} = 0°$.

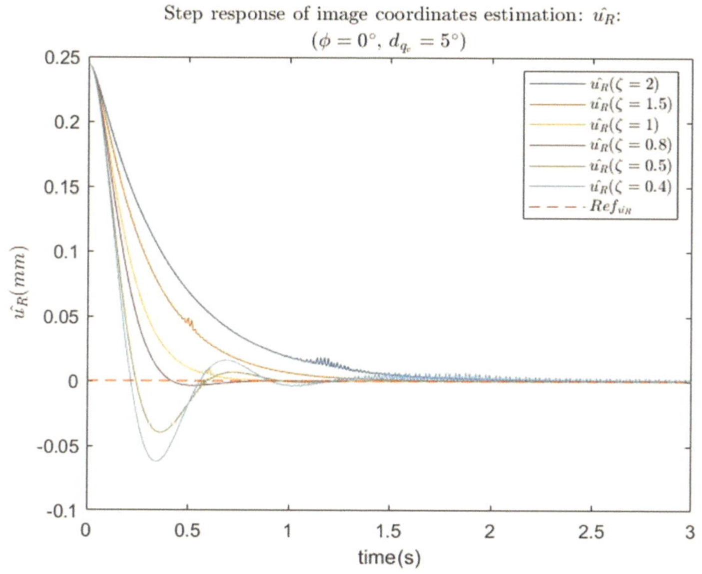

Figure 18.
Step responses of $\hat{u_R}$ *for the case 3:* $\varphi = 0°, d_{q_v} = 5°$.

The case 1 is simulated with the initial angle $\varphi < \frac{\alpha}{2}$, while the case 2 is simulated when $\varphi = \frac{\alpha}{2}$, the largest possible initial angle within the angle of view. It can be shown clearly that without any input disturbance, both methods are able to drive the closed-loop responses to the final value. The step response of the feedback linearization has an overshoot. In addition, for the second method, the model linearization approach, it can be seen from the two simulation cases that there exists a damping ratio, ζ_{opt}, such that.

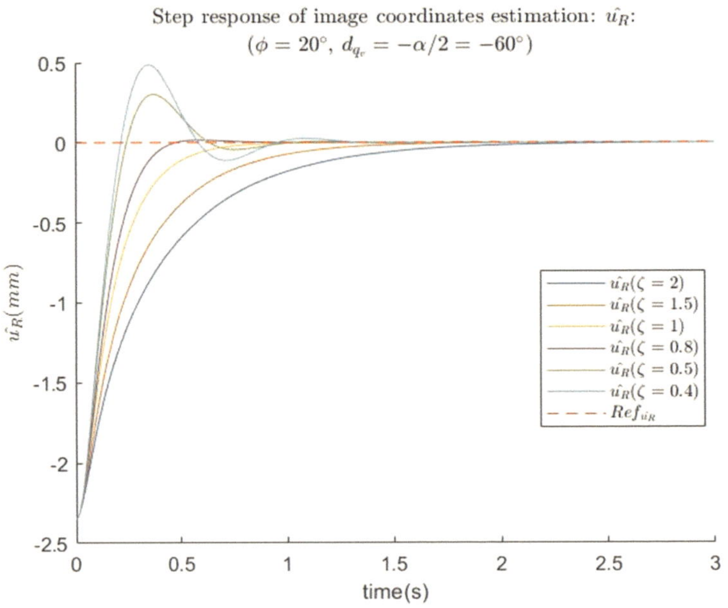

Figure 19.
Step responses of $\hat{u_R}$ *for the case 4:* $\varphi = 20°, d_{q_v} = -\frac{\alpha}{2} = -60°$.

- When $\zeta \geq \zeta_{opt}$, the step responses have no overshoots and as ζ decreases, the system reaches the steady state faster.

- When $\zeta < \zeta_{opt}$, the step responses have overshoots, and the overshoots increase as ζ decreases. As ζ increases, the system reaches the steady state faster.

It can be estimated from **Figures 16** and **17** that $\zeta_{opt} \cong 1$ in the case 1 and $\zeta_{opt} \cong 0.5$ in the case 2. The most desirable system is the one without overshoot and fastest step response. When $\zeta = \zeta_{opt}$, the system has the fastest response and no (or little) overshoot. Therefore, we can state that the best performance of the controlled system is when setting the damping ratio $\zeta = \zeta_{opt}$. Clearly, the value of ζ_{opt} varies with φ, the initial angle of the reference point with respect to the inertial frame.

In the cases 3 and 4, the input disturbance is introduced to the system. In the case 3, a small disturbance ($d_{q_v} = 5°$) is added to the actuator input. The case 4 is a combined case where both φ and d_{q_v} are present ($\varphi = 20°, d_{q_v} = -60°$). **Figures 18** and **19** do not display the step responses of feedback linearization approach. The step responses of feedback linearization are unstable when input disturbances are introduced. It can be shown that any input disturbance drastically alters the nonlinear interface parameters, used in feedback linearization, and hence, results in an unstable system. On the other hand, the linear controller designed based on the linearized plant model is robust to the input disturbances even with the significantly large disturbances (case 4). Similar to the no disturbance cases, ζ_{opt} exists for the cases with disturbances.

From the discussion above, the plant linearization method is the preferred and the recommended method for the camera movement adjustment.

7. The cascaded SISO outer-loop design for high-accuracy robot tool manipulator

The control block diagram of high-accuracy robot tool manipulator is shown in **Figure 8**. In this diagram, the camera is kept static but servos the movement of the tool by the visual data. This block diagram contains a feedback loop as well as a feedforward loop. We design controllers for each loop and simulate the combined loop under different scenarios.

7.1 Developing a combined SISO tool robot arm and camera model

In **Figure 8**, the joint angle including the disturbance of the tool manipulator is transformed to the tool pose using a robot kinematics model. A camera model then is utilized to convert the 3D pose to the 2D, as shown in **Figure 20**. For simplicity, we

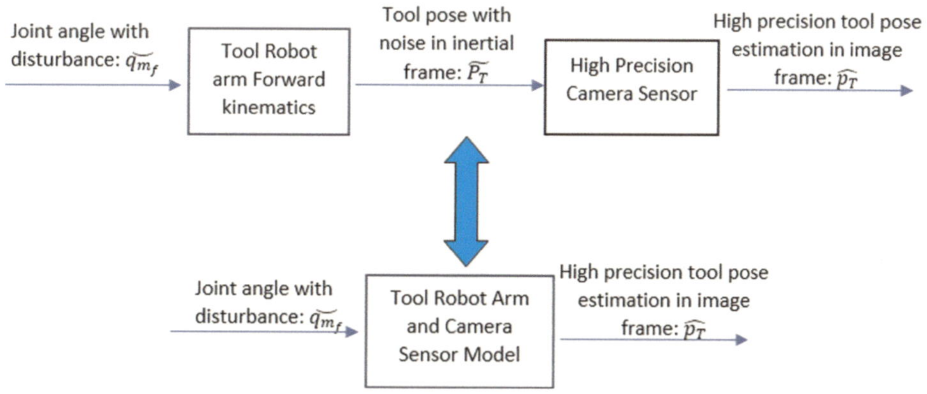

Figure 20.
The tool robot arm and the camera sensor model (a combined block of tool manipulation kinematics and camera sensing).

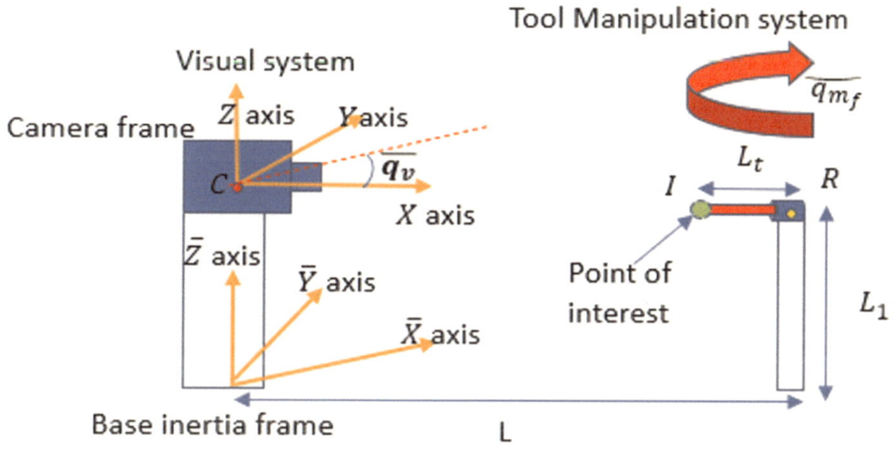

Figure 21.
A SISO camera and tool robot arm setup.

can combine these two blocks into one block, which is called the tool robot arm and camera sensor model.

In **Figure 21**, a SISO combined model setup has been shown based on the one-link camera robot arm model in **Figure 13**. Now, the camera, which is attached to a one-link rotational robot arm captures the image of a tool, which is attached to another similar robot arm, and estimates its angle of rotation. The tool has a length L_t with an interest point is selected at the tip of the tool. Both robot links have a length of L_1 and are separated from each other by a distance L. Assume $\overline{q_v}$ is the angle of the camera from previous control sequences (discussed in **Figure 5**). The inertial and camera coordinate frames set ups are discussed in Section 6.1. The only difference is that the camera frame rotates relative to inertial frame by a clockwise angle $\overline{q_v}$ along the $\overline{Z}$-axis. The tool rotates relative to the $\overline{Z}$-axis in a clockwise direction with a variable angle $\breve{q_m}$. The actual angle of rotation $\breve{q_m}$ is the sum of the input disturbance d_{q_m} and the planned angle of rotation q_m; i.e.,

$$\breve{q_m} = q_m + d_{q_m} \tag{70}$$

Then, we can compute the final angle after rotation by adding the initial angle of the tool in the inertial frame $\boldsymbol{q_{m_0}}$:

$$\breve{q_{mf}} = \breve{q_m} + q_{m_0} \tag{71}$$

The coordinates of the point of interest on the tool in the inertial frame is then computed as $\left(\boldsymbol{L - L_t \cos\left(\breve{q_{mf}}\right), L_t \sin\left(\breve{q_{mf}}\right), L_1}\right)$

Following the same procedures as in Eqs. (37)–(41), we can derive the tool image coordinate $\widehat{\boldsymbol{u_T}}$ along the $\boldsymbol{u}$-axis as:

$$\widehat{u_T} = f_u \frac{Q\left(\breve{q_{mf}}\right) + \tan\left(\overline{q_v}\right)}{1 - Q\left(\breve{q_{mf}}\right) \tan\left(\overline{q_v}\right)} \tag{72}$$

where,

$$Q\left(\breve{q_{mf}}\right) = \frac{L_t \sin\left(\breve{q_{mf}}\right)}{L - L_t \cos\left(\breve{q_{mf}}\right)} \tag{73}$$

Eqs. (72) and (73) provide a function that maps the current or the final angle of the tool onto the image coordinate $\widehat{\boldsymbol{u_T}}$ with constant parameters, $\overline{\boldsymbol{q_v}}$, $\boldsymbol{L}$, and $\boldsymbol{L_t}$.

7.2 The outer-loop feedback and feedforward controller design

The overall plant for the design of this control system is composed of the inner joint control loop, see Eq. (32) and **Figure 8**, and the tool robot arm and camera sensor models, as shown in **Figure 20**. We can design the outer-loop feedback controller using the feedback linearization method or the plant linearization method by following the procedures presented in Section 6.2 and 6.3 respectively. For the sake of brevity, we will not discuss the detail derivations of each controller. Mostly comparative issues such the overshoots and the robustness are discussed in Section 6.4. In this section, we utilize the plant linearization method to design the outer-loop feedback controller.

Without providing the details, the controller is designed for a second order closed-loop system using the plant linearization method (the plant is linearized at $\breve{q}_{mf}$ = 0°) is given as:

$$G_{C_{out}} = \frac{1}{C_1} \frac{(\tau_{in}s+1)^3}{(3\tau_{in}s+1)} \frac{\omega_n{}^2}{s^2+2\zeta\omega_n s} \tag{74}$$

and

$$C_1 = f_u(1+\left(tan\left(\overline{q_v}\right)\right)^2 \frac{LL_t - L_t{}^2}{(L-L_t)^2} \tag{75}$$

Then, the second order closed-loop transfer function $\boldsymbol{T}$ of the overall cascaded control system is expressed as:

$$T = \frac{\omega_n{}^2}{s^2+2\zeta\omega_n s+\omega_n{}^2} \tag{76}$$

Where $\boldsymbol{f_u}$ is the camera focal length, $\overline{\boldsymbol{q_v}}$,$\boldsymbol{L}$, and $\boldsymbol{L_t}$ are the parameters defined in Section 7.1. $\boldsymbol{\tau_{in}}$ defines the bandwidth of the inner joint loop. $\boldsymbol{\omega_n}$ is the natural frequency and $\boldsymbol{\zeta}$ is the damping ratio of the second order system.

As the camera is static in this control stage, the tool pose cannot be recognized and measured visually if it is outside the camera range of view. To tackle this problem, we can estimate the 2D feature (image coordinates of the tool points) from the same model in Eqs. (72) and (73) with the joint angle $\boldsymbol{q_m}$ as input:

$$\tilde{u}_T = f_u \frac{Q(q_m) + \tan\left(\overline{q_v}\right)}{1 - Q(q_m)\tan\left(\overline{q_v}\right)} \tag{77}$$

where

$$Q(q_m) = \frac{L_t \sin(q_m)}{L - L_t \cos(q_m)} \tag{78}$$

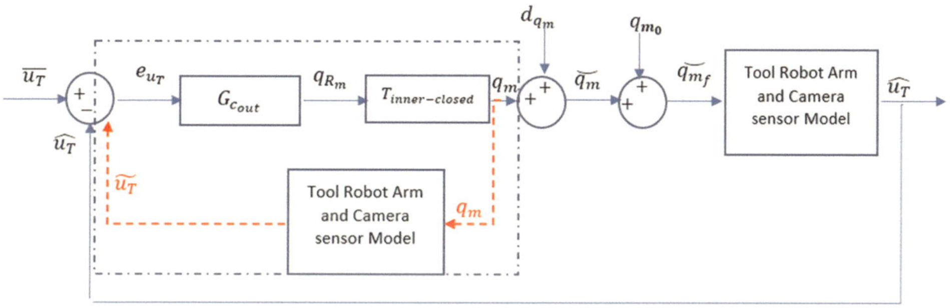

Figure 22.
The block diagram of the tool manipulator feedback control loop with feature estimation.

which is illustrated in the block diagram of **Figure 22**. Normal feedback loop (in blue lines) is preserved when the tool is inside the camera range of view and hence, the camera can estimate the tool 2D feature $\widehat{\boldsymbol{u}_T}$. However, when the tool is outside the range of view, the 2D feature can only be approximated as $\widetilde{\boldsymbol{u}_T}$ (red dashed line) by the combined model as shown in the blue dashed box. We can implement a bump-less switch to smoothly switch between these modes of operations. The switching signal changes over when the tool moves in or out of the camera range of view.

In addition, the feedforward controller, as shown in **Figure 8**, is designed with the inverse kinematics model of the tool robot arm and the details of this design is not provided here. It should be noted, as stated previously, that the combination of the feedforward and the feedback controllers provide a much faster response than the feedback controller by itself.

7.3 The simulation of cascaded SISO closed loop for high-accuracy tool manipulation control system

In this section, we present the simulation results of the feedforward and the feedback control system designed for the robot tool manipulator. As discussed, the plant linearization method is used in the feedback controller design. The controlled system is simulated with the original nonlinear tool robot arm and the camera sensor model of Eqs. (77) and (78). Furthermore, we illustrate how varying the damping ratio ζ affects the responses. We chose the bandwidth of the inner-loop control as $100 rad/s$ and the bandwidth of the outer-loop control as $10 rad/s$. The intrinsic camera parameters are chosen as following: $f_u = 2.8\ mm$ (Focal length) and $\alpha = 120°$ (Angle of view). Other parameters are chosen as: $L = 1m$, and $L_t = 0.135m$.

We compare the performance of the feedforward-feedback control system and the feedback-only control system in two different scenarios. One scenario is simulated when the tool is kept inside the camera angle of view during the entire run time. It should be noted that both the feedback and the feedforward controllers are active during the entire simulation. The second scenario is simulated when the tool is outside the camera angle of view during the entire simulation time. When the tool pose is outside the camera angle of view, the feedback signal is replaced by the estimation signal from the model, as shown in **Figure 22**. Furthermore, for each scenario, we show how varying the damping ratio ζ affects the responses. Assuming the initial angle of the tool, $\boldsymbol{q}_{m_0} = \mathbf{0}$ in the inertial base frame, we vary the pose or the rotational angle of the camera, $\overline{\boldsymbol{q}_v}$, in the inertial base frame, for each simulation scenario.

Figures 23–26 show step responses of the joint angle $\breve{\boldsymbol{q}}_{mf}$ and the output tool pose $\boldsymbol{P}_{TX}$ (only X coordinate of the six dofs pose of the tool $\boldsymbol{P}_T$) for the two scenarios. Each response is simulated with four different damping ratios: ζ=2 (Blue), $\zeta = 1.5$ (Green), $\zeta = 1$ (Purple), and $\zeta = 0.7$ (Black). With the same damping ratio (same color), the response of the feedback-only control system is shown in the dashed line and the response of the feedforward-feedback control system is illustrated in the solid line.

In the first scenario, the camera rotates 5° counterclockwise with respect to the inertial frame. Then, the tool stays in the camera angle of view with any joint angle $\breve{\boldsymbol{q}}_{mf} \in [-\mathbf{180°}, \mathbf{180°}]$. In addition, a disturbance $\boldsymbol{d}_{q_m} = -\mathbf{10°}$, is added to this joint angle. The responses in **Figures 23** and **24** illustrate that both the feedback-only and the feedforward-feedback control systems can reach stability and are robust to the disturbances. With varying the damping ratios, the feedforward-feedback system

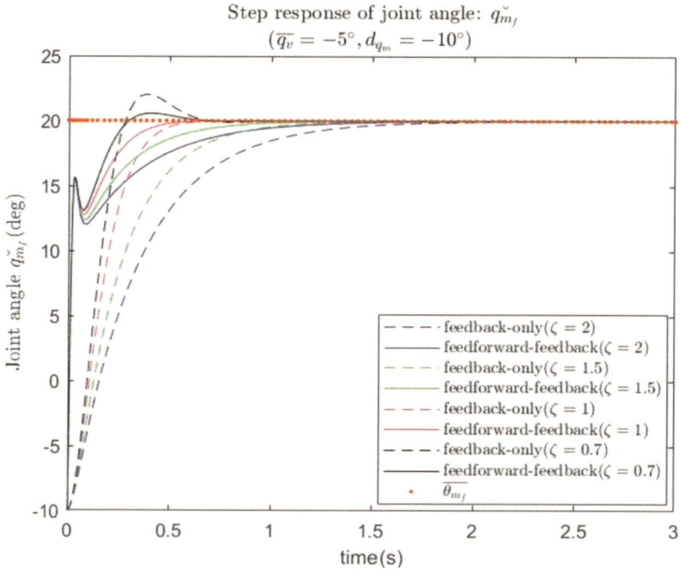

Figure 23.
Step responses of $\breve{q_{mf}}$. *Scenario 1:* $\overline{q_v} = -5°, d_{q_m} = -10°$.

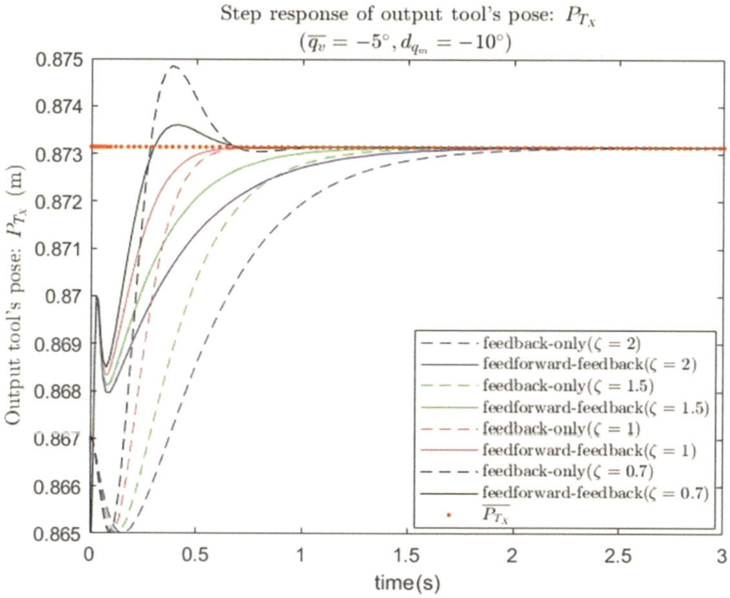

Figure 24.
Step responses of P_{TX}. *Scenario 1:* $\overline{q_v} = -5°, d_{q_m} = -10°$.

responses are faster in transient when compared to the response of the feedback-only system. The feedforward-feedback system response, when the damping ratio is small ($\zeta = 0.7$), results in a less overshoot when compared to the response of the feedback-only system. The optimal damping ratio, $\zeta_{opt} = 1$, (as discussed in Section 6.4), results in the fastest response and no overshoots, both for the feedforward-feedback and the feedback-only control systems.

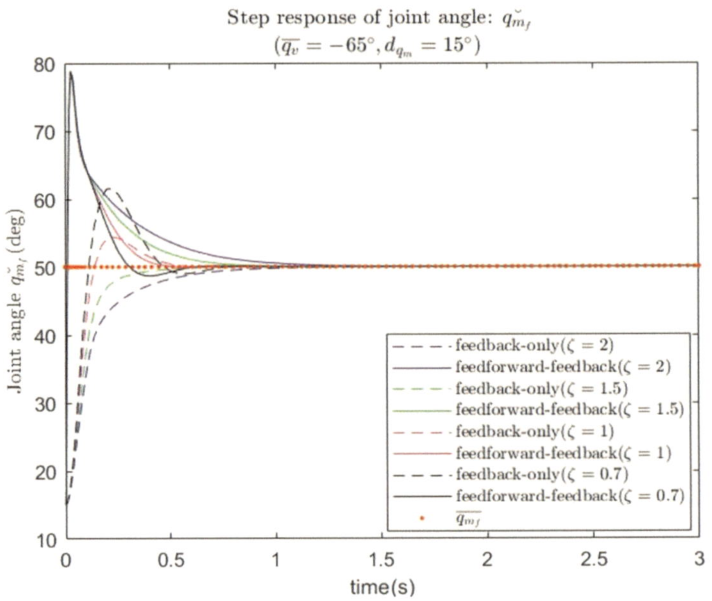

Figure 25.
Step responses of $\breve{q}_{mf}$. *Scenario 2:* $\overline{q_v} = -65°, d_{q_m} = 15°$.

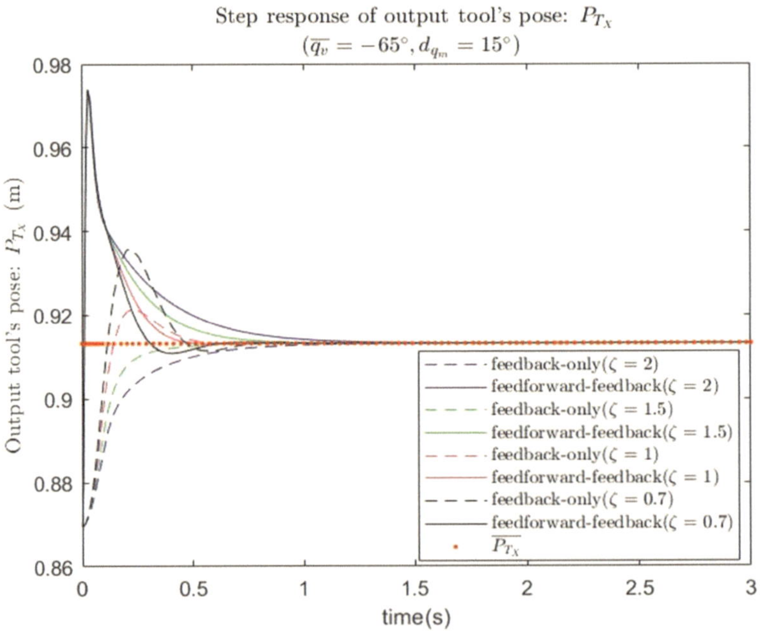

Figure 26.
Step responses of P_{TX}. *Scenario 2:* $\overline{q_v} = -65°, d_{q_m} = 15°$.

In the second scenario, the camera rotates 65° counterclockwise with respect to the inertial frame. Using geometry, we can calculate that the tool is out of the camera range of view when $\breve{q}_{mf} \notin [35.21°, 134.79°]$. Staring from the initial angle $q_{m0} = 0°$

and move to the target joint angle $\overline{q_{mf}} = 50°$, there is a range $\overset{\smile}{q_{mf}} \in [0°, 35.21°]$ that camera cannot detect the tool but estimation of the pose is required to drive the tool to the target. Even perturbed with the disturbance $d_{q_m} = 15°$, all the simulation responses shown in **Figures 25** and **26** reach the target within a second. In this scenario, the feedforward-feedback control system still converges faster in transient but generates bigger overshoots compared to the feedback-only control system. The large overshoots may come from accumulated disturbances that cannot be eliminated by the feedforward control without the intervention of the feedback control. A feedforward controller may drive the tool away from its target even faster when the disturbance appears in the loop. Perhaps, a possible solution, which will be investigated in the future, would be the use of a switching algorithm, switching from a feedforward to a feedback controller, rather than the use of a continuous feedforward-feedback controller.

In summary, the responses from a continuous feedforward-feedback system are more vulnerable to the disturbances especially when the starting position of the tool is far from its target. Although the disturbances will be eliminated as soon as the tool moves inside the camera range of view, the overshoots are more severe if more disturbances are accumulated in the process. In the real-world applications, the feedback-only control solutions are slower than the simulation results as the camera requires extra time, which is not considered in these simulations, to take pictures. Therefore, a feedforward controller, which compensates for the speed limitation of the feedback-only control, becomes indispensable in real manufacturing environments. The issue of the overshoots can be dealt with either by upgrading the camera with a wider range of view or as mentioned previously, the use of a switching algorithm, such as switching from a feedforward to a feedback controller, rather than the use of a continuous feedforward-feedback controller.

8. Conclusion

In this Chapter different sources of uncertainties in the task of positioning control in the automated manufacturing process are introduced. Then, a sequence of control methodologies is proposed. In the first part of this Chapter, we presented movement of a camera in the space to search for an optimal pose, a location in the space where the tool pose can be reached with minimum amount of energy and time duration. In the second part, we discussed a visual servoing architecture, which is applied to eliminate the measurement and dynamic noises occurred in the process of the camera movement. The image averaging technique is used to minimize the image noises by the averaging multiple images. In the last part, we designed the feedback and the feedforward controllers to guide the tool to its target by eliminating the dynamic errors in the tool movement process. Designs of all the control systems have been thoroughly discussed in this Chapter. Our methods for controller design are based on the classical Image Based Visual Servoing (IBVS) technique but are improved on by adding dynamic components to the systems and avoiding the depth estimation as done in the classical methods. Although only one degree of freedom case is discussed in this Chapter, the SISO simulation results have shown great potential of this work for various real-world applications in the automated high-speed manufacturing processes.

Author details

Rongfei Li* and Francis F. Assadian
Department of Mechanical Engineering, University of California, Davis, CA, USA

*Address all correspondence to: rfli@ucdavis.edu

References

[1] Wallén J. The history of the Industrial Robot. Technical Report 2853. Linköping, Sweden: Linköping University, Automatic Control; 2008

[2] Conrad KL, Shiakolas PS, Yih TC. Robotic calibration issues: Accuracy, repeatability and calibration. In: Proceedings of the 8th Mediterranean Conference on Control and Automation (MED2000); 17–19 July 2000. Rio, Patras, Greece: Springer; 2000

[3] Mironov D, Altamirano M, Zabihifar H, Liviniuk A, Liviniuk V, Tsetserukou D. Haptics of Screwing and Unscrewing for Its Application in Smart Factories for Disassembly. Cha Springer; 2018. DOI: 10.1007/978-3-319-93399-3_37

[4] Anonymous. Robotic Accuracy Improves Aerospace Manufacturing. Industry Automation Review. 2020. Available fro https://industrialautomationreview.com/robotic-accuracy-improves-aerospace-manufacturing/ [Accessed: November 30, 2021]

[5] Chaumette F, Hutchinson S. Visual servo control. I. Basic approaches. IEEE Robotics & Automation Magazine. 2006; **13**(4):82-90. DOI: 10.1109/MRA.2006.250573

[6] Hutchinson SA, Hager GD, Corke PI. A tutorial on visual servo control. IEEE Transactions on Robotics and Automation. 1996;**12**(5):651-670. DOI: 10.1109/70.538972

[7] Chaumette F, Hutchinson S. Visual servoing and visual tracking. In: Siciliano B, Oussama K, editors. Handbook of Robotics. Berlin Heidelberg, Germany: Springer-Verlag; 2008. pp. 563-583. DOI: 10.1007/978-3-540-30301-5.ch25

[8] Wilson WJ, Hulls CCW, Bell GS. Relative end-effector control using cartesian position based visual servoing. IEEE Transactions on Robotics and Automation. 1996;**12**(5):684-696. DOI: 10.1109/70.538974

[9] Youla D, Jabr H, Bongiorno J. Modern wiener-Hopf Design of Optimal Controllers-Part II: The multivariable case. IEEE Transactions on Automatic Control. 1976;**21**(3):319-338. DOI: 10.1109/TAC.1976.1101223

[10] Zhu W, Liu H, Ke Y. Sensor-based control using an image point and distance features for rivet-in-hole insertion. IEEE Transactions on Industrial Electronics. 2019;**67**(6): 4692-4699. DOI: 10.1109/TIE.2019.2927186

[11] Liu H, Zhu W, Ke Y. Pose alignment of aircraft structures with distance sensors and CCD cameras. Robotics and Computer-Integrated Manufacturing. 2017;**48**:30-38. DOI: 10.1016/j.rcim.2017.02.003

[12] Liu H, Zhu W, Dong H, Ke Y. An adaptive ball-head positioning visual servoing method for aircraft digital assembly. Assembly Automation. 2019; **39**(2):287-296. DOI: 10.1108/AA-05-2018-066

[13] Du G, Wang K, Lian S, Zhao K. Vision-based robotic grasping from object localization, object pose estimation to grasp estimation for parallel grippers: A review. Artificial Intelligence Review. 2021;**54**:1677-1734. DOI: 10.1007/s10462-020-09888-5

[14] Avigal Y, Paradis S, Zhang H. 6-DoF grasp planning using fast 3D reconstruction and grasp quality CNN.

Machine Learning Papers. 2020; **8**:1-6. DOI: 10.48550/arXiv.2009.08618.

[15] Wu C, Chen J, Cao Q, Zhang J, Tai Y, Sun L, et al. Grasp proposal networks: an end-to-end solution for visual learning of robotic grasps. Advances in Neural Information Processing Systems. 2020; **33**:13174-13184. DOI: 10.48550/arXiv.2009.12606

[16] Li R, Platt R, Yuan W, Pas A, Roscup N, Srinivasan M, et al. Localization and manipulation of small parts using GelSight tactile sensing. IThe IEEE/RSJ International Conference on Intelligent Robots and Systems (IROS). 2014;**20**:3988-3993. DOI: 10.1109/IROS.2014.6943123

[17] Dong S, Rodriguez A. Tactile-based insertion for dense boxpacking. The IEEE/RSJ International Conference on Intelligent Robots and Systems (IROS). 2019;**21**:7953-7960. DOI: 10.1109/IROS40897.2019.8968204

[18] Hogan F, Ballester J, Dong S, Rodriguez A. Tactile dexterity: Manipulation primitives with tactile feedback. The IEEE International Conference on Robotics and Automation (ICRA). 2020;**23**:8863-8869. DOI: 10.1109/ICRA40945.2020

[19] Fazeli N, Oller M, Wu J, Wu Z, Tenenbaum J, Rodriguez A. See, feel, act: hierarchical learning for complex manipulation skills with multisensory fusion. Science Robotics. 2019;**4**(26): eaav3123. DOI: 10.1126/scirobotics.aav3123

[20] Gregorio D, Zanella R, Palli G, Pirozzi S, Melchiorri C. Integration of robotic vision and tactile sensing for wire-terminal insertion tasks. IEEE Transactions on Automation Science and Engineering. 2019;**16**(2):585-598

[21] Flandin G, Chaumette F, Marchand E. Eye-in-hand/eye-to-hand cooperation for visual Servoing. Proceedings—IEEE International Conference on Robotics and Automation. 2000;**3**(3):2741-2746. DOI: 10.1109/ROBOT.2000.846442

[22] Hartley R, Zisserman A. Multiple View Geometry in Computer Vision. 2nd ed. Cambridge: Cambridge University Press; 2004. DOI: 10.1017/CBO9780511811685

[23] Sonka M, Hlavac V, Boyle R. Image Processing, Analysis, and Machine Vision. 3rd ed. Pacific Grove: Brooks/Cole Publishing Company; 1999. pp. 373-421. DOI: 10.1007/978-1-4899-3216-.ch9

[24] Zhang Z. A flexible new technique for camera calibration. IEEE Transactions on Pattern Analysis and Machine Intelligence. 2000;**22**(11): 1330-1334. DOI: 10.1109/34.888718

[25] Tsai R. A versatile camera calibration technique for high-accuracy 3D machine vision metrology using off-the-shelf TV cameras and lenses. IEEE Journal on Robotics and Automation. 1987;**3**(4): 323-344. DOI: 10.1109/JRA.1987.1087109

[26] Hartley. An algorithm for self-calibration from several views. In: Proceedings of IEEE Conference on Computer Vision and Pattern Recognition. London: IEEE; 1994. pp. 908-912. DOI: 10.1109/CVPR.1994.323923

[27] Luong QT, Faugeras O. Self-calibration of a moving camera from point correspondences and fundamental matrices. International Journal of Computer Vision. 1997;**22**:261-289. DOI: 10.1023/A:1007982716991

[28] Mark WS, MV. Robot Dynamics and Control. Hoboken, New Jersey, United States: John Wiley & Sons, Inc.; 1989

[29] Anonymous. ABB IRB 4600-40/2.55 Product Manual. 2013. Available from: https://www.manualslib.com/manual/1449302/Abb-Irb-4600-40-2-55.html#manual [Accessed: December 01, 2021]

[30] Denavit J, Hartenberg RS. A kinematic notation for lower-pair mechanisms based on matrices. Transactions of the ASME, Journal of Applied Mechanics. 1955;**23**(2):215-221. DOI: 10.1115/1.4011045

[31] Elatta AY, Gen LP, Zhi FL, Daoyuan Y, Fei L. An overview of robot calibration. Information Technology Journal. 2004;**3**:74-78. DOI: 10.3923/itj.2004.74.78

[32] Renders J, Rossignol E, Becquet M, Hanus R. Kinematic calibration and geometrical parameter identification for robots. IEEE Transactions on Robotics and Automation. 1991;**7**(6):721-732. DOI: 10.1109/70.105381

[33] Hayati SA. Robot arm geometric link parameter estimation. In: The 22nd IEEE Conference on Decision and Control. London: IEEE; 1983. pp. 1477-1483. DOI: 10.1109/CDC.1983.269783

[34] Zhuang H, Wang L, Roth ZS. Simultaneous calibration of a robot and a hand-mounted camera. Proceedings IEEE International Conference on Robotics and Automation. 1993;**2**:149-154. DOI: 10.1109/ROBOT.1993.292139

[35] Wu C. A kinematic CAD tool for the design and control of a robot manipulator. The International Journal of Robotics Research. 1984;**3**(1):58-67. DOI: 10.1177/027836498400300105

[36] Mooring BW, Roth ZS, Driels MR. Fundamental of Robotic Calibration. John Willey and Sons; 1991. pp. 221-225

[37] Mooring BW, Tang GR. An improved method for identifying the kinematic parameters in a six axis robot. In: Proceedings of the ASME Computers in Engineering Conference, (ACES'83). Las Vegas, Nevada: ACES; 1984. pp. 79-84

[38] Tsai RY, Lenz RK. A new technique for fully autonomous and efficient 3D robotics hand/eye calibration. IEEE Transactions on Robotics and Automation. 1989;**5**(3):345-358. DOI: 10.1109/70.34770

[39] Mooring BW, Padavala SS. The effect of kinematic model complexity on manipulator accuracy. Proceedings of International Conference on Robotics and Automation. 1989;**1**:593-598. DOI: 10.1109/ROBOT.1989.100049

[40] Khalil W, Gautier M, Enguehard C. Identifiable parameters and optimum configurations for robots calibration. Robotica. 1991;**9**(1):63-70. DOI: 10.1017/S0263574700015575

[41] Lee K, Shah DK. Dynamic analysis of a three-degrees-of-freedom in-parallel actuated manipulator. IEEE Journal on Robotics and Automation. 1988;**4**(3):361-367. DOI: 10.1109/56.797

[42] Patidar P, Gupta M, Srivastava S, Nagawat AK. Image De-noising by various filters for different noise. International Journal of Computer Applications. 2010;**9**:45-50

[43] Das S, Saikia J, Das S, Goñi N. A comparative study of different noise filtering techniques in digital images. International Journal of Engineering Research and General Science. 2015;**3**(5):180-191

[44] Zhao R, Cui H. Improved threshold denoising method based on wavelet transform. In: 2015 7th International Conference on Modelling, Identification and Control (ICMIC). London: IEEE; 2015. pp. 1-4. DOI: 10.1109/ICMIC.2015.7409352

[45] Ng J, Goldberger JJ. Signal averaging for noise reduction. In: Goldberger J, Ng J, editors. Practical Signal and Image Processing in Clinical Cardiology. London: Springer; 2010. pp. 69-77. DOI: 10.1007/978-1-84882-515-4.ch7

[46] Fischler MA, Bolles RC. Random sample consensus: A Paradigm for model fitting with applications to image analysis and automated cartography. In: Fischler MA, Firschein O, editors. Readings in Computer Vision. San Francisco: Morgan Kaufmann; 1987. pp. 726-740. DOI: 10.1016/B978-0-08-051581-6.50070-2

[47] Chen G, Zhu F, Heng PA. An efficient statistical method for image noise level estimation. In: 2015 IEEE International Conference on Computer Vision (ICCV). London: IEEE; 2015. pp. 477-485. DOI: 10.1109/ICCV.2015.62

[48] Assadian F, Mallon K. Robust Control: Youla Parameterization Approach. Hoboken, New Jersey, USA: Jon Wiley & Sons, Inc.; 2022

Chapter 8

Robust Control Based on Input-Output Feedback Linearization for Induction Motor Drive: Real Time Implementation

Saber Krim and Mohamed Faouzi Mimouni

Abstract

This chapter proposes a design of hardware architecture of an improved Direct Torque Control (DTC) for a real-time implementation on a Xilinx Field-Programmable Gate Array (FPGA). The first contribution in this chapter consists in combining the DTC with a Space Vector Modulation (SVM) technique and an Input-Output Feedback Linearization (IOFL) approach. In fact, the classical DTC has remarkable performance in terms of fast torque response and less dependence on the system parameters. Despite the cited advantages, the classical DTC is penalized by high torque ripples and inverter-switching-frequency variations. In this context, the SVM is added to the DTC structure in order to keep the switching frequency constant and to reduce ripples. Furthermore, the nonlinear IOFL is proposed to achieve a decoupled flux and torque control. The novel structure is named in this chapter as DTC-IOFL-SVM. Moreover, this chapter presents a hardware implementation of the suggested DTC-IOFL-SVM strategy utilization. The hardware implementation is chosen in order to reduce the sampling period of the system thanks to the parallel processing of the FPGA. In order to demonstrate the performance of the FPGA implementation of the proposed DTC-IOFL-SVM, numerous simulation results are presented using the Xilinx system generator under a Matlab/Simulink.

Keywords: induction motor, direct torque control, input–output feedback linearization, FPGA

1. Introduction

Recently, the Direct Torque Control (DTC) of electrical machines has taken the attention of several researchers, thanks to its interest and advantages, like the simple structure, the fast torque response, and the less dependence on machine parameters [1–3]. The structure of the classical DTC is mainly based on two hysteresis controllers and a lookup table to independently control the torque and the flux by selecting the optimal voltage vector in each sampling period. The classical DTC suffers from several problems like the torque ripples, the harmonics in the stator current waves, as well as the variation in the switching frequency. The fixed bands of the hysteresis controllers are the main cause of these problems [4, 5].

In the recent years, several methods have been put forward for overcoming the classical DTC problems, such as the use of intelligent techniques like the artificial neural networks and the fuzzy logic [6, 7]. However, the experimental implementation of the intelligent techniques requires powerful calculation processes due to their complexity. The torque and flux ripples, and the stator current waveform distortions can be reduced by fixing the switching frequency and selecting the more appropriate voltage vector for each commutation period. Indeed, in order to impose an operation with a fixed switching frequency, a combination between the DTC and the Space Vector Modulation (SVM) has been proposed by several research studies [8–10]. In fact, the DTC with a fixed switching frequency consists in introducing two Proportional Integral PI controllers and a SVM technique to achieve the best choice of the voltage vector in each sampling period [8–10]. However, the stability and dynamics of the system will be affected by the variation in machine parameters due to the existence of the PI controllers. In order to get rid of the drawbacks of the mentioned techniques, several robust control techniques have been proposed in order to guarantee the high performance control of induction motor drives. Among of these techniques we can cite the sliding mode control, the backstepping control and the Input–Output Feedback Linearization (IOFL) approach [11–13], which are the most popular control strategies. IOFL consists in transforming a nonlinear system into an equivalent linear one, which can be utilized for controlling the system [14]. IOFL is based on an inverse mathematical transformation for obtaining a suitable control law of the Induction Motor (IM).

The main first objective of this chapter consists in combining the IOFL technique and an SVM-DTC (SVM-DTC-IOFL) in order to design a novel DTC strategy featured by fast torque and speed responses, more robustness under stator resistance variations, reduced ripples and distortions, and a decoupled control between the torque and the flux. In this study, the stator flux and the electromagnetic torque are chosen as control states to develop the decoupled model of the IM.

For real time control of electrical machines, digital electronic boards like the STM32-microcontrollers [15, 16] and the Digital Signal Processor (DSP) are usually utilized [17–20]. The digital circuits based on microprocessors are known by their sequential computation of the control algorithm which consequently increases the execution time and the sampling period when the complexity of the control algorithm increases. Indeed, if the sampling time raises, the delays in the control system goes up, this causes additional ripples and distortions in the torque and the current, respectively. Moreover, the DSP controllers are chosen for implanting the control algorithms of electrical systems [21, 22], which are based on processor cores with high performance and few peripherals to communicate with the external environment. In fact, the sampling period of the processor depends of the computational burden due to the parallel processing, which creates delays in the feedback loop and raises the stator current harmonics and the torque ripples [23–25].

With the target of overcoming the DSP limitation and minimizing the DSP computational burden, a combination between the DSP and the FPGA has been proposed in the literature [26, 27] with the purpose of distributing the computational burden between these two digital controllers. This solution offers better performance by reducing the sampling period, the ripples in the torque and the distortions in the stator current.

However, the main limitations of this solution are the high cost and the complexity of circuit's connections, which causes problems for commercialization. In order to overcome the limitations of the cited solutions, the FPGA can be used only for controlling the motor drives. Indeed, thanks to its hardware architecture, the FPGA offers good performance by reducing the execution time and consequently the delays in the retroaction loop. In the last few years, the DSP (DSPACE 1104) has been suggested and confirmed by several engineers and researchers for real time control of AC machines [24, 28, 29]. In the same context, the FPGA can overcome the software solution drawbacks by adopting parallel processing [30–33]. In fact, the FPGA offers the designer the possibility of implementing in a low sampling period, control techniques with good performance and high algorithmic complexities. Indeed, in [32], the authors have implemented a control algorithm of an IM using an FPGA under a sampling period of 5 μs [32].

The second objective of this chapter consists in implementing the proposed SVM-DTC-IOFL on an FPGA board. For the hardware implementation on the FPGA, the SVM-DTC-IOFL must be transformed into VHDL or Verilog description languages. Indeed, VHDL or Verilog programming is a difficult task which raises the design time, the time to market and the system cost. In this chapter, a graphical programming method based on Xilinx System Generator (XSG) is utilized in order to reduce the prototyping time. In fact, the graphical architecture from the XSG under a Matlab/Simulink-tool makes it possible to generate the VHDL of the Verilog code, as well as the programming bitstream files [33–35]. The XSG is a toolbox created by the Xilinx engineers' team, which operates between Matlab and Vivado tools, whose objective is to facilitate the programming tasks and reduce the time to market [35].

In this chapter, SVM-DTC-IOFL is theoretically developed, designed from the XSG tool, and verified by digital simulation utilizing a Xilinx Zynq FPGA.

This work is composed of five sections. In Section 2, the state mode of an induction motor drive, the SVM technique principle and the suggested IOFL theory are presented. In Section 3, designs from the XSG of the proposed SVM-DTC-IOFL and simulation results are shown. The implementation and synthesis results are given in Section 4. The conclusion is summarized in Section 5.

2. Theory and modeling

In this chapter, a combination between the SVM technique, the DTC strategy and the IOFL technique is put forward. The SVM is suggested in order to prevent ripples and distortions, and it provides an operation with a fixed switching frequency. IOFL is used in order to achieve decoupled control between the torque and flux quantities. The principle of these techniques is detailed in the following subsections.

2.1 Model presentation

The IM model is presented as follows, which will be used to design the proposed IOFL approach.

$$\dot{x} = f(x) + g\, v_{s\alpha\beta} \tag{1}$$

with:

$$x = \begin{bmatrix} i_{s\alpha} & i_{s\beta} & \phi_{s\alpha} & \phi_{s\beta} \end{bmatrix}^T \tag{2}$$

$$f(x) = \begin{bmatrix} -\frac{1}{\sigma}\left(\frac{1}{T_r}+\frac{1}{T_s}\right) i_{s\alpha} - \omega_m i_{s\alpha} + \frac{1}{\sigma L_s T_r}\phi_{s\alpha} + \frac{\omega_m}{\sigma L_s}\phi_{s\beta} \\ \omega_m i_{s\alpha} - \frac{1}{\sigma}\left(\frac{1}{T_r}+\frac{1}{T_s}\right) i_{s\beta} - \frac{\omega_m}{\sigma L_s}\phi_{s\alpha} + \frac{1}{\sigma L_s T_r}\phi_{s\beta} \\ -R_s i_{s\alpha} \\ -R_s i_{s\beta} \end{bmatrix} \quad (3)$$

$$g = \begin{bmatrix} \frac{1}{\sigma L_s} & 0 & 1 & 0 \\ 0 & \frac{1}{\sigma L_s} & 0 & 1 \end{bmatrix} \quad (4)$$

where:

$(i_{s\alpha}, i_{s\beta})$:the stator current components,

$(v_{s\alpha}, v_{s\beta})$:the voltage vectors components,

$(\phi_{s\alpha}, \phi_{s\beta})$:the stator flux vector components,

(R_r, R_s) : the rotor and stator resistance respectively,

(L_r, L_s) : the rotor and stator inductance respectively,

(T_r, T_s) : the rotor and stator time constants,

ω_m (red/sec): the electric rotor speed.

$\sigma = 1 - \frac{M_{sr}^2}{L_r L_s}$: The Blondel coefficient, where M_{sr} presents the mutual inductance.

2.2 Space vector modulation

The classical DTC based on fixed-bandwidth hysteresis controllers produces high ripples and distortions. Indeed, if a larger hysteresis-band of the torque is chosen, the torque ripples increase. For a smaller hysteresis band, the torque ripples are reduced and the switching frequency goes up, which consequently increases the commutation losses in the inverter IGBT transistors [36]. Thus, the SVM technique is proposed in this chapter in order to maintain a fixed switching frequency and reduce the ripples [37, 38]. The SVM principle consists in modulating reference voltage vector components in order to generate the more appropriate voltage vector that characterizes

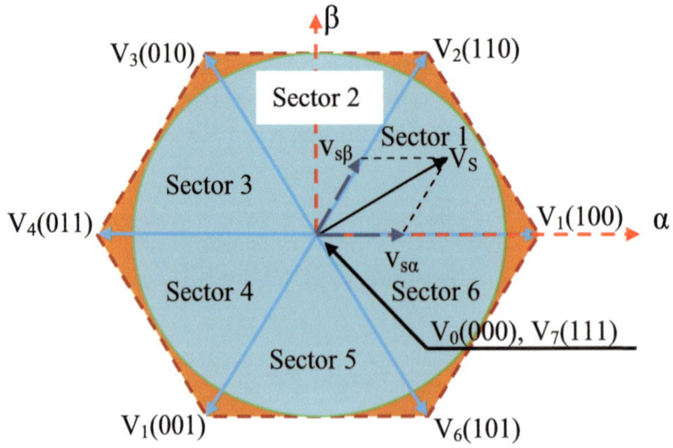

Figure 1.
Voltage vectors.

inverter control signals. As shown in **Figure 1**, the reference voltage vector can be determined by projecting it on the two vectors that bound the sector, using Eq. (5).

The time allowed for each voltage vector application can be determined by vector calculations. The rest of the sampling period can be filled by applying the null vector in order to grantee a fixed switching frequency [39, 40]. An example for the first section, by projection on vectors V_1 and V_2, the voltage vector application times T_1 and T_2 are given by Eq. (5):

$$\begin{cases} \vec{V}_S = V_{s\alpha}^* + jV_{s\beta}^* = \dfrac{T_1}{T_m}\vec{V}_1 + \dfrac{T_2}{T_m}\vec{V}_2 \\ \vec{V}_1 = \sqrt{\dfrac{2}{3}}U_{dc}(\cos(0) + j\sin(0)) = \sqrt{\dfrac{2}{3}}U_{dc} \\ \vec{V}_2 = \sqrt{\dfrac{2}{3}}U_{dc}\left(\cos\left(\dfrac{\pi}{3}\right) + j\sin\left(\dfrac{\pi}{3}\right)\right) \\ T_m = T_1 + T_2 + T_0 \\ T_1 = \left(\sqrt{\dfrac{3}{2}}V_{S\alpha} - \dfrac{1}{\sqrt{2}}V_{S\beta}\right)\dfrac{T_m}{U_{dc}} \\ T_2 = \sqrt{2}V_{S\alpha}\dfrac{T_{mod}}{U_{dc}} \end{cases} \tag{5}$$

where $\left(v_{s\alpha}^*, v_{s\beta}^*\right)$ represents the components of the reference voltage vector, T_1 and T_2 denote the commutation time, T_m is the sampling time, and U_{dc} is the DC voltage.

2.3 IOFL theory

This section illustrates the Feedback Linearization (FL) based DTC for an IM drive. The FL technique utilizes an inverse mathematical transformation in order to determine the desired control law for controlling the nonlinear system such as the IM. Furthermore, the FL technique is utilized to obtain decoupled control between the torque and flux. In this study, the suggested system outputs are the electromagnetic torque and the square root of the stator flux norm. Referring to the IOFL theory, the output variables are expressed as:

$$\begin{cases} h_1(x) = T_{em} = \dfrac{3}{2}N_p\left(i_{s\beta}\phi_{s\alpha} - i_{s\alpha}\phi_{s\beta}\right) \\ h_2(x) = |\phi_s|^2 = \phi_{s\alpha}^2 + \phi_{s\beta}^2 \end{cases} \tag{6}$$

where T_{em} is the estimated electromagnetic torque, and $|\phi_s|$ is the norm of the stator flux. Assuming the controller objectives y_1 and y_2 as, we get:

$$\begin{cases} y_1 = h_1(x) - T_{em}^* = T_{em} - T_{em}^* \\ y_2 = h_2(x) - |\phi_s^*|^2 = |\phi_s|^2 - |\phi_s^*|^2 \end{cases} \tag{7}$$

where T_{em}^* and $|\phi_s^*|$ are the torque and flux references, respectively. Utilizing the presented equations, the time derivative of the controller objectives can be written as:

$$\begin{bmatrix} \dot{y}_1 \\ \dot{y}_2 \end{bmatrix} = \begin{bmatrix} g_1(x) \\ g_2(x) \end{bmatrix} + G(x) \begin{bmatrix} v_{s\alpha} \\ v_{s\beta} \end{bmatrix} \tag{8}$$

with:

$$\begin{cases} g_1(x) = \frac{3}{2} N_p \left[-\frac{1}{\sigma} \left(\frac{1}{T_r} + \frac{1}{T_s} \right) \phi_{s\alpha} i_{s\alpha} + \omega_m \phi_{s\alpha} i_{s\alpha} - \frac{\omega_m}{\sigma L_s} \phi_{s\alpha}^2 \right. \\ \left. \qquad + \frac{1}{\sigma} \left(\frac{1}{T_r} + \frac{1}{T_s} \right) \phi_{s\beta} i_{s\alpha} + \omega_m \phi_{s\beta} i_{s\beta} - \frac{\omega_m}{\sigma L_s} \phi_{s\beta}^2 \right] - \dot{T}_{em}^* \\ g_2(x) = -2R_s \phi_{s\alpha} i_{s\alpha} - 2R_s \phi_{s\beta} i_{s\beta} - \left| \dot{\phi}_s^* \right| \\ G(x) = \begin{bmatrix} 2\phi_{s\alpha} & 2\phi_{s\beta} \\ \frac{3}{2} N_p \left(i_{s\beta} - \frac{1}{\sigma L_s} \phi_{s\beta} \right) & \frac{3}{2} N_p \left(i_{s\alpha} - \frac{1}{\sigma L_s} \phi_{s\alpha} \right) \end{bmatrix} \end{cases} \tag{9}$$

Based on the IOFL technique, the control inputs can be expressed as follows [41].

$$\begin{bmatrix} v_{s\alpha} \\ v_{s\beta} \end{bmatrix} = G^{-1}(x) \begin{bmatrix} -g_1(x) + v_1 \\ -g_2(x) + v_2 \end{bmatrix} \tag{10}$$

where v_1 and v_2 are assumed to be two auxiliary inputs with the purpose of ensuring more desired behavior and tracking accuracy for the torque and the stator flux, with:

$$\begin{cases} v_1 = -k_1 y_1 \\ v_2 = -k_2 y_2 \end{cases} \tag{11}$$

where k_1 and k_2 are positive constants. The SVM-DTC-IOFL performance strongly depends on the suitable choice of parameters k_1 and k_2. In fact, the high values of such parameters are able to cause the system instability. On the other hand, the small values will lead to a poor robustness and slow convergence. Finally, it is necessary to better choose such parameters for guarantying high control technique performance [13]. The combination between (8), (10) and (11) gives the following expression:

$$\begin{bmatrix} \dot{y}_1 \\ \dot{y}_2 \end{bmatrix} = \begin{bmatrix} -k_1 & 0 \\ 0 & -k_2 \end{bmatrix} \begin{bmatrix} y_1 \\ y_2 \end{bmatrix} \tag{12}$$

Utilizing the IM model, the relation between the rotor and the stator fluxes is given below:

$$\begin{cases} \phi_{r\alpha} = \frac{\sigma L_s L_r}{M_{sr}} \left(\frac{1}{\sigma L_s} \phi_{s\alpha} - i_{s\alpha} \right) \\ \phi_{r\beta} = \frac{\sigma L_s L_r}{M_{sr}} \left(\frac{1}{\sigma L_s} \phi_{s\beta} - i_{s\beta} \right) \end{cases} \tag{13}$$

Utilizing matrix G(x), defined in (9) and Eq. (13), the determinant of G(x) is given as follows:

$$G(x) = \frac{3M_{sr}}{\sigma L_s L_r} N_p \left(\phi_{r\alpha} \phi_{s\alpha} + \phi_{r\beta} \phi_{s\beta} \right) \tag{14}$$

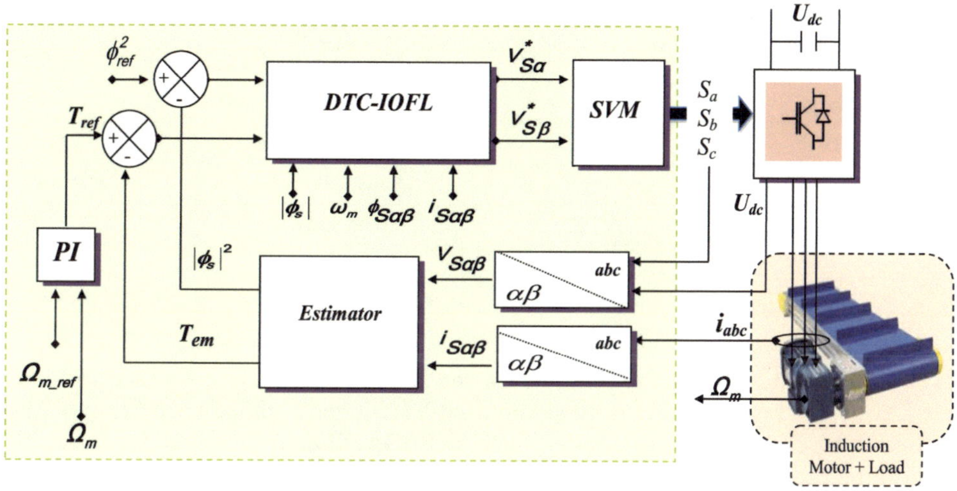

Figure 2.
Global diagram of the proposed SVM-DTC-IOFL.

Referring to Eq. (14), it can be noticed that the product between the rotor flux and the stator flux cannot be zero, and matrix G(x) is nonsingular [42].

The FL control law is used in order to satisfy the stability condition defined by the Lyapunov approach. To study the stability of the control law, the Lyapunov function is given as:

$$V = \frac{1}{2} y^T y \tag{15}$$

The time derivative of (15) is given as follows:

$$\dot{V} = y^T \dot{y} = \begin{bmatrix} y_1 & y_2 \end{bmatrix} \begin{bmatrix} -k_1 & 0 \\ 0 & -k_2 \end{bmatrix} \begin{bmatrix} y_1 \\ y_2 \end{bmatrix} = -k_1 y_1^2 - k_2 y_2^2 < 0 \tag{16}$$

Parameters k_1 and k_2 are positive, so derivative $\dot{V}$ is negative, which demonstrates the stability of the control system. The global diagram of the proposed SVM-DTC-IOFL is given by **Figure 2**.

3. Simulation results and discussion

In this section, the simulation studies of an IM controlled by two control strategies, named classical DTC and SVM-DTC-IOFL, have been carried out under a Matlab/Simulink environment. The hardware architecture of the two control strategies are designed using XSG tool. The different parameters of the IM model are provided in **Table 1**.

The XSG tool is developed by Xilinx to be integrated into a Matlab/Simulink environment. It is widely utilized for the design, verification and implementation of control algorithms architectures on FPGAs. When we get the desired design with good

Parameter	Value	Parameter	Value
power (kW)	1.5	Rotor resistance (Ω)	4.282
Voltage (V)	230/400	Stator inductance (H)	0.464
Frequency (Hz)	50	Rotor inductance (H)	0.464
Pole pair	2	Mutual inductance (H)	0.4417
Stator resistance (Ω)	5.717	Rated speed (rpm)	1435

Table 1.
Induction machine parameters.

of simulation results, it will be possible for the XSG to automatically generate the VHDL code. As a matter of fact, the generated VHDL code will be used for generating the download. Bit file to be integrated into the FPGA. **Figure 3** depicts the design flow through the use of the XSG. **Figure 4** presents the SVM-DTC-IOFL architecture from the XSG.

3.1 First scenario

In this scenario, a comparative study between the classical DTC and the proposed SVM-DTC-IOFL is done under a rated load torque (10 Nm), a variable speed profile

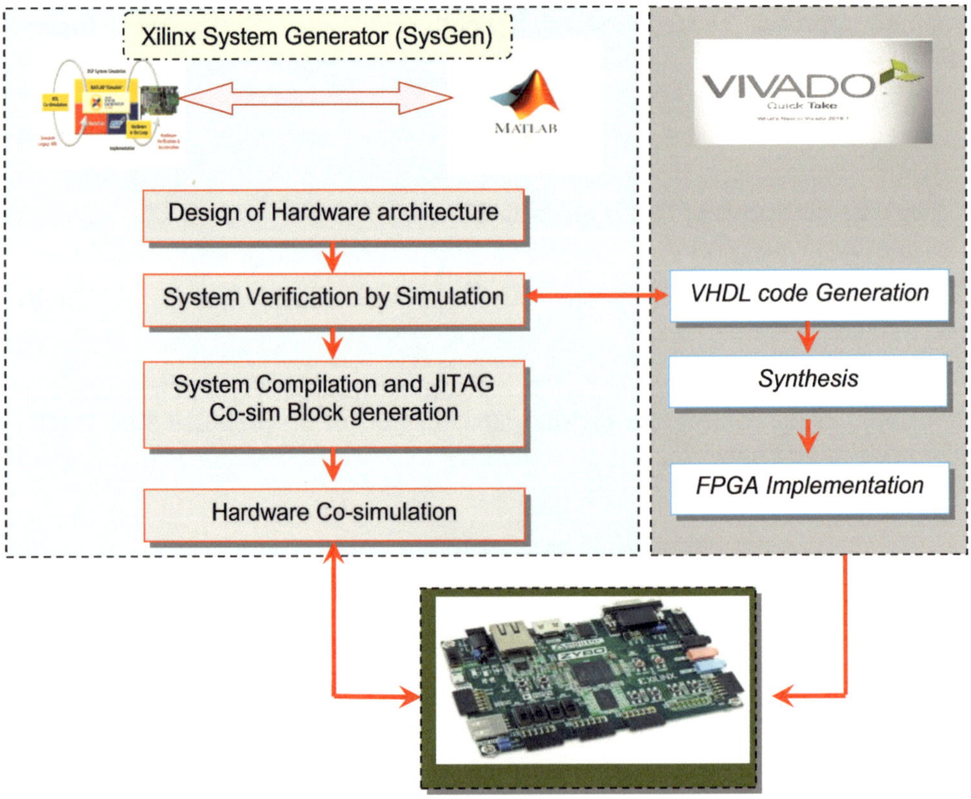

Figure 3.
XSG design flow.

and a reversal of the direction of rotation. In order to show the effectiveness of the suggested SVM-DTC-IOFL, it is compared with the classical DTC in terms of torque ripples and stator current distortion. The performance analysis is carried out with a sampling period equal to 100 μs.

The IM starts with a reference speed equal to 100 rad/sec. At t = 1 sec the reference speed decreases slowly to reach −100 rad/sec at t = 2 sec. At t = 0.5 sec, a rated torque is applied.

Figure 5 presents the evolution of the rotor speed of the IM controlled by two control strategies. It can be noticed that the rotor speed converges to the reference speed for both control strategies. However, the proposed SVM-DTC-IOFL offers better performance in terms of ripples around the reference speed, as shown in **Figure 5(b)**. As given in **Figure 6(a)**, the suggested control strategy gives better performance in terms of ripples compared to the classical DTC (**Figure 6(b)**). **Figure 7** presents the three phase stator current consumed by the IM control by both control strategies. It can be seen that the proposed control strategy offers better performance in terms current distortions. In fact, for the suggested SVM-DTC-IOFL, the stator current has a smooth waveform (**Figure 7(a)**). **Figure 8** presents the evolution of the stator flux components for both control strategies. In can be seen that

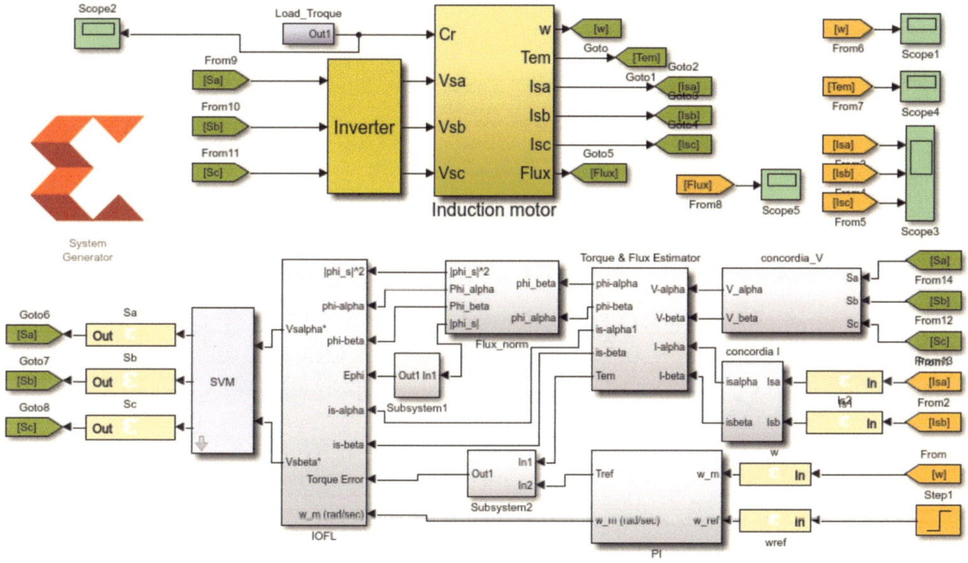

Figure 4.
Full architecture of SVM-DTC-IOFL using XSG.

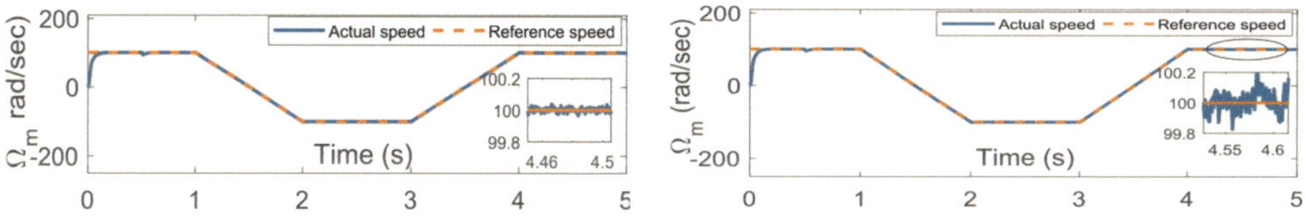

(a) (b)

Figure 5.
Speed response for: (a) proposed SVM-DTC-IOFL, (b) classical DTC.

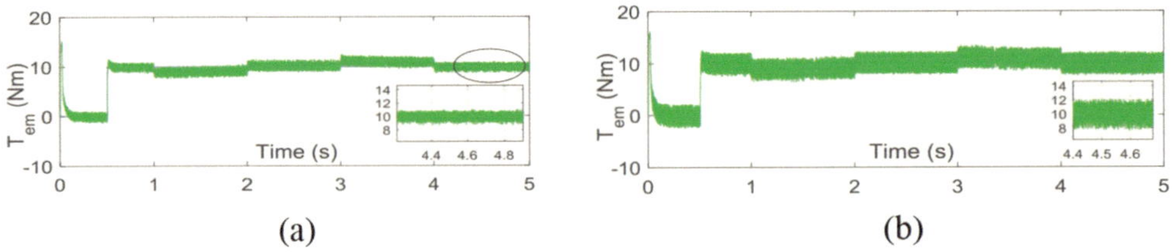

Figure 6.
Torque response for: (a) proposed SVM-DTC-IOFL, (b) classical DTC.

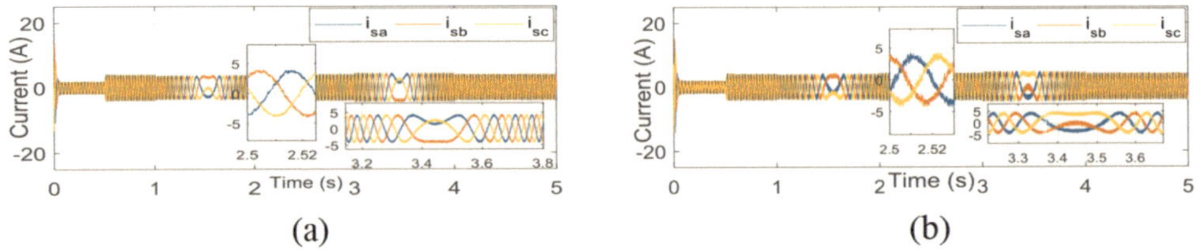

Figure 7.
Three phase stator current for: (a) proposed SVM-DTC-IOFL, (b) classical DTC.

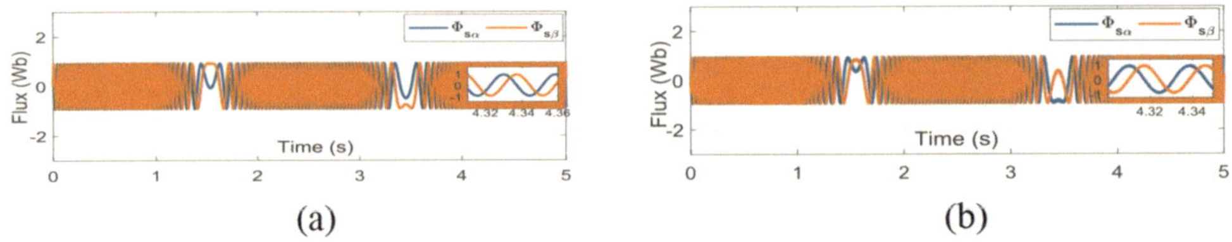

Figure 8.
Three phase stator current for: (a) suggested SVM-DTC-IOFL, (b) classical DTC.

the real stator flux converges to its reference value. In addition, the proposed control strategy gives better performance in terms of flux-ripple reduction. More details are illustrated in **Table 2**.

3.2 Second scenario

In this scenario we used the same simulation conditions of the first scenario, but the main deference consists in reducing the sampling period which is equal to 10 μs. In fact, when the control algorithm is implemented on software solutions like the

	Classical DTC	Proposed SVM-DTC-IOFL
Speed ripples	Medium	neglected
Torque ripples (%)	40%	20%
Current distortion	High	Medium
Sampling period	100 μs	100 μs

Table 2.
Comparison between the both control strategies.

microcontrollers or the DSP, the sampling time increased due to the serial processing of these solutions, which consequently raises the control loop delay, the torque ripples and the stator current distortions. In order to overcome the limitations of these solutions in terms of execution time, the FPGA is proposed thanks to its parallel processing and short execution time. In order to show the effects of the execution time on the simulation results, a sampling period of 10 μs is chosen. The obtained results in this scenario demonstrate that when the sampling period decreases, the torque and the stator flux ripples, as well as the stator current harmonics, are reduced, as shown in **Tables 2** and **3**.

The IM starts with a reference speed equal to 100 rad/sec. At t = 1 sec, the reference speed falls slowly to reach −100 rad/sec at t = 2 sec. At t = 0.5 sec, a rated torque is applied.

Figure 9 depicts the evolution of the rotor speed of the IM controlled by two control strategies. It can be noticed that the rotor speed converges to the reference speed for both control strategies. However, the suggested SVM-DTC-IOFL offers better performance in terms of ripples around the reference speed, as shown in **Figure 9(b)**. As given by **Figure 10(a)**, the proposed control strategy provides better performance in terms of ripples compared to the classical DTC (**Figure 10(b)**). **Figure 11** presents the three phase stator current consumed by the IM control by both control strategies. It can be seen that the suggested control strategy offers better performance in terms current distortions. In fact, for the proposed SVM-DTC-IOFL, the stator current has a smooth waveform (**Figure 11(a)**). **Figure 12** presents the evolution of the extremity of the stator flux vector in the Concordia reference. It can be noticed that when the motor is controlled by the classical DTC, the stator flux vector trajectory presents high deviations and ripples (as shown by **Figure 12(b)**). Contrariwise, in the case of the proposed SVM-DTC-IOFL a smooth circular trajectory is obtained as illustrated in **Figure 12(a)**. More details are given in **Table 3**.

	Classical DTC	Proposed SVM-DTC-IOFL
Speed ripples	Medium	neglected
Torque ripples (%)	10%	5%
Flux ripples (%)	4.39%	1.09%
Current distortion	High	neagleted
Sampling period	10 μs	10 μs

Table 3.
Comparison between the both control strategies.

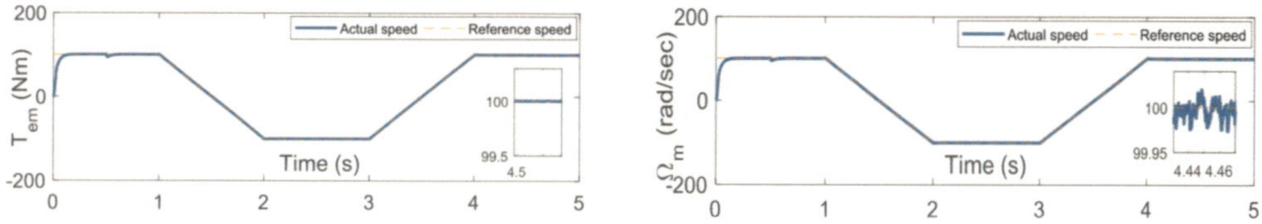

(a) (b)

Figure 9.
Speed response for: (a) proposed SVM-DTC-IOFL, (b) classical DTC.

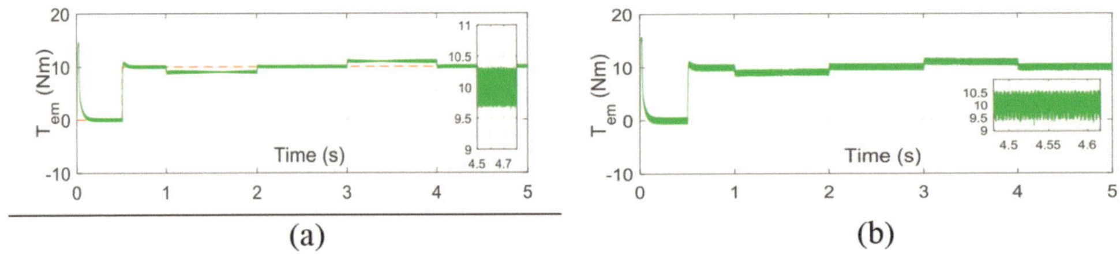

Figure 10.
Torque response for: (a) proposed SVM-DTC-IOFL, (b) classical DTC.

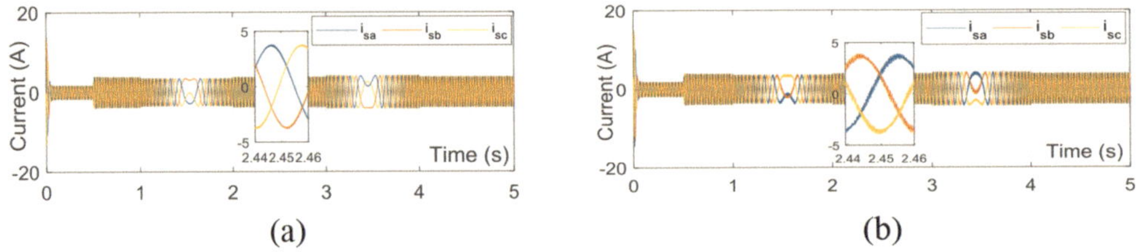

Figure 11.
Three phase stator current for: (a) proposed SVM-DTC-IOFL, (b) classical DTC.

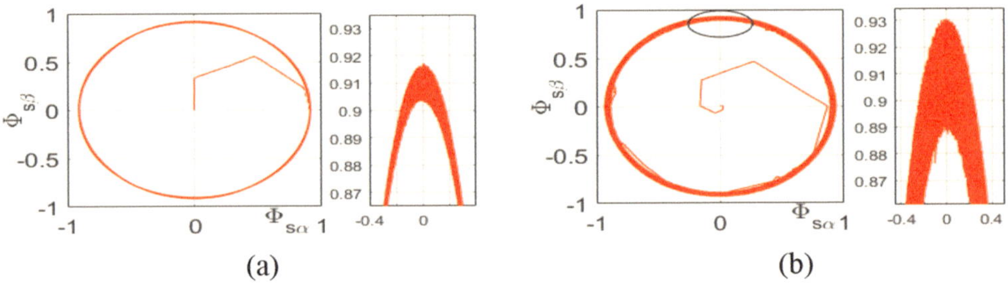

Figure 12.
Three phase stator flux for: (a) proposed SVM-DTC-IOFL, (b) classical DTC.

3.3 Third scenario

This section consists in testing the robustness of the proposed SVM-DTC-IOFL under stator resistance variations at a low speed region. In this study, the IM starts with a reference speed equal to 20 rad/sec. The sampling period is equal to 10 μs. At t = 4 sec, the stator resistance increases to reach 1.5 R_{sn}. **Figure 13(a, b)**, presents the evolution of the rotor speed for both control strategies. As shown in **Figure 13(a)**, it can be seen that the suggested SVM-DTC-IOFL offers better performance with a small deviation when the stator resistance goes up.

Figure 14(a, b) illustrates the evolution of the stator flux module for both control strategies. Referring to **Figure 14(a)**, it can be noticed that when the stator resistance rises, the stator flux curve presents small deviations and then it converges quickly to its reference value. However, when the IM is controlled by the classical DTC, the actual stator flux diverges from its reference value due to the variation in the stator resistance.

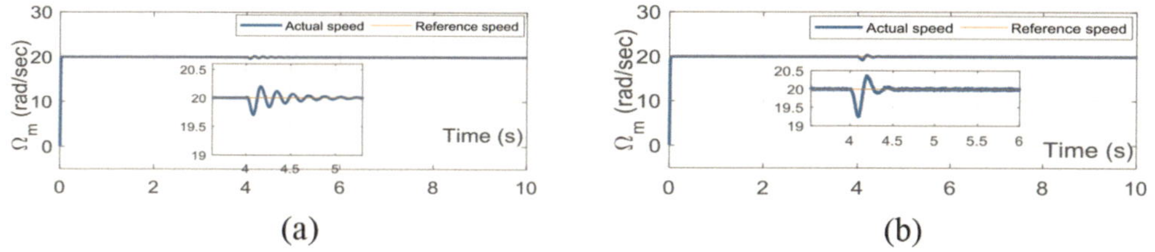

Figure 13.
Speed response for: (a) proposed SVM-DTC-IOFL, (b) classical DTC.

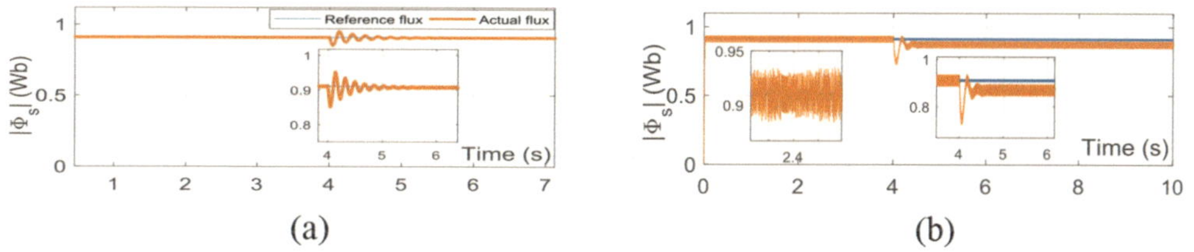

Figure 14.
Speed response for: (a) proposed SVM-DTC-IOFL, (b) classical DTC.

4. VHDL code generation and synthesize results

The VHDL code generation and synthesis steps can be validated after verifying the functionality of the XSG architecture of the proposed SVM-DTC-IOFL. The obtained simulation results of the section confirm the good functionality of the designed XSG architecture, which offers the possibility to generate the VHDL and determine the synthesis results utilizing the Xilinx Vivado. During the hardware implementation of the classical DTC and the proposed SVM-DTC-IOFL approaches, the used resources from the FPGA are depicted in **Table 4**.

	Used with SVM-DTC-IOFL	Available
LUT	1436 (2.7%)	53,200
LUTRAM	104 (0.59%)	17,400
FF	2826 (2.65%)	106,400
BUFG	1 (3.12%)	32
DSP	10 (4%)	220

Table 4.
Utilized resources from Xilinx Zynq FPGA.

5. Conclusion

In this chapter, a performance improvement of the DTC of an IM drive utilizing the SVM technique and a nonlinear control technique named IOFL has been presented. In order to solve the classical DTC problems, like the torque ripples, the current distortion and the variation in the switching frequency, the SVM has been developed in this chapter. The proposed scheme is known as SVM-DTC. To increase the robustness of the suggested scheme under parameter variations, an IOFL approach has been combined with the SVM-DTC to generate the reference voltage vector. The real time implementation on the Xilinx Zynq FPGA has been put forward and investigated in this chapter so as to reduce the period of the system and eliminate the time delay in the control loop. The design of the proposed scheme has been carried out using the XSG toolbox. The flux and torque ripples have been considerably reduced thanks to the SVM technique. The nonlinear approach has given more performance, such as the robustness against the parameter variations, good and fast dynamic response and good tracking, and has reduced the complexity of the control scheme. Furthermore, the designed architecture of the control algorithm has been tested with two different sampling periods in order to demonstrate that if the sampling period rises, the ripples increase. Moreover, this controller has been featured by its simple design and implementation. The hardware FPGA implementation of the proposed SVM-DTC-IOFL can be considered as a good solution to control electrical motor drives.

For future work we are interested in the experimental validation of the proposed DTC-SVM-IOFL utilizing a real test bench.

Conflict of interest

The authors declare no conflict of interest.

Author details

Saber Krim[1,2*] and Mohamed Faouzi Mimouni[2,3]

1 Higher Institute of Applied Sciences and Technology of Kasserine, University of Kairouan, Kairouan, Tunisia

2 Laboratory of Automatic, Electrical Systems and Environment, National Engineering School of Monastir, University of Monastir, Tunisia

3 National Engineering School of Monastir, University of Monastir, Monastir, Tunisia

*Address all correspondence to: krim.saber.enim@gmail.com

References

[1] Casadei D, Profumo F, Serra G, Tani A. FOC and DTC: Two viable schemes for induction motors torque control. IEEE Transactions on Power Electronics. 2002;**17**:779-787

[2] Ren Y, Zhu ZQ. Enhancement of steady-state performance indirect-torque- controlled dual three-phase permanent-magnet synchronous machine drives with modified switching table. IEEE Transactions on Industrial Electronics. 2015;**62**:3338-3350

[3] Lai Y, Wang W, Chen Y. Novel switching techniques for reducing the speed ripple of AC drives with direct torque control. IEEE Trans on Industrial Electronics. 2004;**51**(4):768-775

[4] Kang J-W, Sul SK. Analysis and prediction of inverter switching frequency in direct torque control of induction machine based on hysteresis bands and machine parameters. IEEE Transactions on Industrial Electronics. 2001;**48**:545-553

[5] Idris NRN, Yatim AHM. Direct torque control of induction machines with constant switching frequency and reduced torque ripple. IEEE Transactions on Industrial Electronics. 2004;**51**(4)

[6] Saberi H, Feyzi M, Sharifian MBB, Sabahi M. Improved sensorless direct torque control method using adaptive flux observer. IET Power Electronics. 2014;**7**:1675-1684

[7] Fu X, Li S. A novel neural network vector control technique for induction motor drive. IEEE Transactions on Energy Conversion. 2015;**30**:1428-1437

[8] Rashag HF, Tan NML, Koh SP, Abdalla AN, Chong KH, Tiong SK. DTC-SVM based on PI torque and PI flux controllers to achieve high performance of induction motor. Research Journal of Applied Sciences, Engineering and Technology. 2014;**7**(4):875-891

[9] Joseline Metilda A, Arunadevi R, Ramesh N, Sharmeela C. Analysis of direct torque control using space vector modulation for three phase induction motor. Recent Research in Science and Technology. 2011;**3**(7):37-40

[10] Saber KRIM, Gdaim S, Mtibaa A, Mimouni MF. Implementation on the FPGA of DTC-SVM based proportional integral and sliding mode controllers of an induction motor: A comparative study. Journal of Circuits, Systems, and Computers. 2017;**26**(3)

[11] Zaafouri A, Ben C, Ben AH, Châari A. DSP-based adaptive backstepping using the tracking errors for high-performance sensorless speed control of induction motor drive. ISA Transactions. 2015:1-15

[12] Orlowska-Kowalska T, Tarchala G, Dybkowski M. Sliding-mode direct torque control and sliding-mode observer with a magnetizing reactance estimator for the field-weakening of the induction motor drive. Mathematics and Computers in Simulation. 2014:31-45

[13] Alonge F, Cirrincione M, Pucci M, Sferlazza A. Input–output feedback linearization control with online MRAS-based inductor resistance estimation of linear induction motors including the dynamic end effects. IEEE Transactions on Industry Applications. 2016:254-266

[14] Yazdanpanah R, Soltani J, Arab Markadeh GR. Nonlinear torque and stator flux controller for induction motor drive based on adaptive input–output feedback linearization and sliding mode

control. Energy Conversion and Management. 2008;**49**:541-550

[15] Motor Control With STM32®32-Bit ARM®-Based MCU. Pdf. Available from: http://www.st.com/web/en/resource/sales_and_marketing/promotion al_material/brochure/brstm32mc.pdf

[16] STM32 embedded target for MATLAB and Simulink, Available from: http://www.st.com/st-web-ui/static/active/en/resource /technical/ document/ data_brief/DM00080897.pdf

[17] Ozcira S, Bekiroglu N, Senol I. Dynamic performance and analysis of direct torque control method based on DSP for PMSM drives. In: Proceedings of the IEEE International Conference on Renewable Energy Research and Applications; Nagasaki, Japan. 2012. pp. 11-14

[18] Lianbing L, Hexu S, Xiaojun W, Yongqing T. A high-performance direct torque control based on DSP in permanent magnet synchronous motor drive. In: Proceedings of the IEEE 4th World Congress on Intelligent Control and Automation (Cat. No. 02EX527); Shanghai, China. Vol. 2. 10-14 June 2002. pp. 1622-1625

[19] Cruz SMA et al. DSP implementation of the multiple reference frames theory for the diagnosis of stator faults in a DTC induction motor drive. IEEE Transactions on Energy Conversion. 2005;**20**:329-335

[20] Singh B, Singh BP, Dwivedi S. DSP based implementation of sliding mode speed controller for direct torque controlled PMSM drive. In: Proceedings of the IEEE International Conference on Industrial Technology; Mumbai, India. 2006. pp. 1301-1308

[21] Monmasson E, Cirstea MN. FPGA design methodology for industrial control systems—A review. IEEE Transactions on Industrial Electronics. 2007;**54**(4):1824-1842

[22] Cecati C. Microprocessors for power electronics and electrical drives applications. IEEE Ind. Electron. Soc. Newslett. 1999;**46**(3):5-9

[23] Hussein F, Malik E. Improving the torque ripple in DTC of PMSM using fuzzy logic. In: IEEE Industry Applications Society Annual Meeting; Edmonton, AB, Canada. 5-9 October 2008. pp. 1-8. DOI: 10.1109/08IAS.2008.162

[24] Idkhajine L, Monmasson E, Naouar MW, Prata A, Bouallaga K. Fully integrated FPGA-based controller for synchronous motor drive. IEEE Transactions on Industrial Electronics. 2009;**56**(10)

[25] Lokriti A, IssamSalhi SD. IM direct torque control with no flux distortion and no static torque error. ISA Transactions. 2015;**59**:256-267

[26] Idris NRN, Chuen LT, Elbuluk ME. A new torque and flux controller for direct torque control of induction machines. IEEE Transactions on Industry Applications. 2006;**42**(6):1358-1366

[27] Jidin A, Idris NRN, Yatim AHM, Sutikno T, Elbuluk ME. An optimized switching strategy for quick dynamic torque control in DTC-hysteresis-based induction machines. IEEE Transactions on Industrial Electronics. 2011;**58**(8): 3391-3400

[28] Tabbache B, Benbouzid M, Kheloui A, Bourgeot JM, Mamoune A. An improved fault-tolerant control scheme for PWM inverter-fed induction motor-based EVs. ISA Transactions. 2013;**52**:862-869

[29] Rafa S, Larabi A, Barazane L, Manceur M, Essounbouli N,

Hamzaoui A. Implementation of a new fuzzy vector control of induction motor. ISA Transactions. 2014;**52**:744-754

[30] Shahbazi M, Poure P, Saadate S, Zolghadri MR. FPGA-based reconfigurable control for fault-tolerant Back-to-Back converter without redundancy. IEEE Transactions on Industrial Electronics. 2013;**60**(8)

[31] Jezernik K, Korelic J, Horvat R. PMSM sliding mode FPGA based control for torque ripple reduction. IEEE Transactions on Power Electronics. 2013;**28**(7)

[32] Sutikno T, Idris NR, Jidin A, Cirstea MN. An improved FPGA implementation of direct torque control for induction machines. IEEE Transactions on Industrial Informatics. 2013;**9**(3)

[33] Gafsi M, Abbassi N, Hajjaji MA, Malek J, Mtibaa A. Xilinx Zynq FPGA for hardware implementation of a chaos-based cryptosystem for real-time image protection. Journal of Circuits, Systems and Computers. 2021;**30**(11): 2150204

[34] Saidani T, Atri M, Dia D, Tourki R. Using Xilinx system generator for real time hardware Co-simulation of video processing system. Electronic Engineering and Computing Technology. 2010;**60**:227-236

[35] Xilinx System Generator User's Guide. UG640 (v11.4), 2009. Available from: http://www.xilinx.com/support/documentation/sw_manuals/xilinx11/sysgen_user.pdf

[36] Uddin M, Hafeez M. FLC-based DTC scheme to improve the dynamic performance of an IM drive. IEEE Transactions on Industry Applications. 2012;**48**:823-831. DOI: 10.1109/TIA.2011.2181287

[37] Bounadja M, Belarbi A, Belmadani B. A high performance space vector modulation–direct torque controlled induction machine drive based on stator flux orientation technique. Adv. Electr. Comput. Eng. 2009;**9**(2):28-33

[38] Rashag HF, Koh SP, Chong KH, Tiong SK, Tan NML, Abdalla AN. High performance of space vector modulation direct torque control SVM-DTC based on amplitude voltage and stator flux angle. Research Journal of Applied Sciences, Engineering and Technology. 2013;**5**(15): 3934-3940

[39] Habetler TG, Profumo F, Pastorelli M, Tolbert LM. Direct torque control of induction machines using space vector modulation. Ind Appl IEEE Trans. 1992;**28**:1045-1053. DOI: 10.1109/28.158828

[40] Ammar A, Bourek A, Benakcha A. Modified load angle direct torque control for sensorless induction motor using sliding mode flux observer. In: Proceedings of the IEEE 4th Conference on Electrical Engineering; Boumerdes, Algeria. 13-15 December 2015. pp. 1-6

[41] Soltani J, Markadeh GRA, Abjadi NR, Ping HW. A new adaptive direct torque control (DTC) scheme based-on SVM for adjustable speed sensorless induction motor drive. In: Proceedings of the IEEE International Conference on Electrical Machines and Systems (ICEMS); Seoul, Korea (South). 8-11 October 2007. pp. 1111-1116

[42] Zhang Z, Tang R, Bai B, Xie D. Novel direct torque control based on space vector modulation with adaptive stator flux observer for induction motors. IEEE Transactions on Magnetics. 2010;**46**: 3133-3136